An Electrochemical Reactor to Minimize Brominated DBPs in a Conventional Treatment Plant

Awwa Research Foundation

About the Awwa Research Foundation

The Awwa Research Foundation (AwwaRF) is a member-supported, international, nonprofit organization that sponsors research to enable water utilities, public health agencies, and other professionals to provide safe and affordable drinking water to consumers.

The Foundation's mission is to advance the science of water to improve the quality of life. To achieve this mission, the Foundation sponsors studies on all aspects of drinking water, including supply and resources, treatment, monitoring and analysis, distribution, management, and health effects. Funding for research is provided primarily by subscription payments from approximately 1,000 utilities, consulting firms, and manufacturers in North America and abroad. Additional funding comes from collaborative partnerships with other national and international organizations, allowing for resources to be leveraged, expertise to be shared, and broad-based knowledge to be developed and disseminated. Government funding serves as a third source of research dollars.

From its headquarters in Denver, Colorado, the Foundation's staff directs and supports the efforts of more than 800 volunteers who serve on the board of trustees and various committees. These volunteers represent many facets of the water industry, and contribute their expertise to select and monitor research studies that benefit the entire drinking water community.

The results of research are disseminated through a number of channels, including reports, the Web site, conferences, and periodicals.

For subscribers, the Foundation serves as a cooperative program in which water suppliers unite to pool their resources. By applying Foundation research findings, these water suppliers can save substantial costs and stay on the leading edge of drinking water science and technology. Since its inception, AwwaRF has supplied the water community with more than $300 million in applied research.

More information about the Foundation and how to become a subscriber is available on the Web at **www.awwarf.org.**

An Electrochemical Reactor to Minimize Brominated DBPs in a Conventional Treatment Plant

Prepared by:
Lina Boulos, **Hélène Baribeau**, **Brian Carrico**, and **Gil Crozes**
Carollo Engineers, P.C., Pasadena, CA 91101

David Kimbrough
Castaic Lake Water Agency

and

Mel Suffet
University of California, Los Angeles

Jointly sponsored by:
Awwa Research Foundation
6666 West Quincy Avenue, Denver, CO 80235-3098

and

Castaic Lake Water Agency
27234 Bouquet Canyon Road, Santa Clarita, California 91350

Published by:

Awwa Research Foundation

Distributed by:

American Water Works Association

IWA Publishing

DISCLAIMER

This study was jointly funded by the Awwa Research Foundation (AwwaRF) and Castaic Lake Water Agency (CLWA). AwwaRF and CLWA assume no responsibility for the content of the research study reported in this publication or for the opinions or statements of fact expressed in the report. The mention of trade names for commercial products does not represent or imply the approval or endorsement of AwwaRF or CLWA. This report is presented solely for informational purposes.

CONTENTS

TABLES

FIGURES

FOREWORD

The Awwa Research Foundation (AwwaRF) is a nonprofit corporation that is dedicated to the implementation of a research effort to help utilities respond to regulatory requirements and traditional high-priority concerns of the industry. The research agenda is developed through a process of consultation with subscribers and drinking water professionals. Under the umbrella of a Strategic Research Plan, the Research Advisory Council prioritizes the suggested projects based upon current and future needs, applicability, and past work; the recommendations are forwarded to the Board of Trustees for final selection. The foundation also sponsors research projects through the unsolicited proposal process; the Collaborative Research, Research Applications, and Tailored Collaboration programs; and various joint research efforts with organizations such as the U.S. Environmental Protection Agency, the U.S. Bureau of Reclamation, and the Association of California Water Agencies.

This publication is a result of one of these sponsored studies, and it is hoped that its findings will be applied in communities throughout the world. The following report serves not only as a means of communicating the results of the water industry's centralized research program but also as a tool to enlist the further support of the nonmember utilities and individuals.

Projects are managed closely from their inception to the final report by the foundation's staff and large cadre of volunteers who willingly contribute their time and expertise. The foundation serves a planning and management function and awards contracts to other institutions such as water utilities, universities, and engineering firms. The funding for this research effort comes primarily from the Subscription Program, through which water utilities subscribe to the research program and make an annual payment proportionate to the volume of water they deliver and consultants and manufacturers subscribe based on their annual billings. The program offers a cost-effective and fair method for funding research in the public interest.

A broad spectrum of water supply issues is addressed by the foundation's research agenda: resources, treatment and operations, distribution and storage, water quality and analysis, toxicology, economics, and management. The ultimate purpose of the coordinated effort is to assist water suppliers to provide the highest possible quality of water economically and reliably. The true benefits are realized when the results are implemented at the utility level. The foundation's trustees are pleased to offer this publication as a contribution toward that end.

David E. Rager
Chair, Board of Trustees
Awwa Research Foundation

Robert C. Renner, P.E.
Executive Director
Awwa Research Foundation

ACKNOWLEDGMENTS

First and foremost, the authors would like to acknowledge Castaic Lake Water Agency for providing in-kind laboratory analyses and experimental runs in the later stages of the project. We would especially like to acknowledge the laboratory staff of Castaic Lake Water Agency, namely Dr. Yao Kouame for anions analyses, Howard An for THM analyses, Parvaneh Moheban for TOC, HPC, and general minerals, and Majid Lanroodi for metals. A special thank you to Khoa Ly for running the high-power reactor experiments as in-kind to this project.

The support from AwwaRF Project Manager, Dr. Kenan Ozekin, was truly appreciated. The authors would also like to thank the support from Roy Martinez, for expediting the contract amendment during the Tailored Collaboration Review process.

The authors would like to thank the Project Advisory Committee members, Dr. Paul Westerhoff (Arizona State University), Dr. Mark Williams (Williams-McCaron, Inc.), and Craig Thompson (Kennedy/Jenks), for their practical input, and a special thank you to Dr. Paul Westerhoff and his student Hye Jin Moon at Arizona State University for providing one series of in-kind ozonation work on raw and electrolyzed samples.

Thank you to the technical advisors on this project: Dr Gary Christian (University of Washington) for his expertise in electrochemistry principles and mechanisms, Dr. Rhodes Trussell (Trussell Technologies) for bridging the gap between electrochemistry and practical water treatment issues, Dr. Susan Springthorpe from the University of Ottawa for providing her expertise on microbial inactivation by electrolysis, and Dr. Rick Sakaji formerly from California Department of Public Health for his input on regulatory issues.

More thanks to the two student interns Alex Revchuk (UCLA-Environmental Science and Engineering Program) and Cynthia Thung (part-time UCLA student and Carollo intern at the time of the study) for running the reactor experiments and multiple analyses at the laboratory.

The authors would like to thank Dr. Graham Juby and Dr. Adam Zacheis, from Carollo Engineers, for their input on scaling-up issues. The authors would also like to acknowledge Eric Hoopes from Intuitech for his input in the reactor reconfiguration.

Finally, the authors would like to acknowledge the technical assistance of Stacy Fuller and Stacey Lara for the preparation of this manuscript, and the assistance of Jenifer Martinez for miscellaneous administrative tasks.

EXECUTIVE SUMMARY

Meeting the United Stated Environmental Protection Agency (USEPA) disinfectant and disinfection by-product (D/DBP) rule has proved challenging for water utilities as they are required to decrease microbial risk with disinfectants, while simultaneously reduce the chemical risk caused by disinfection by-products (DBP). The presence of bromide in drinking water supplies complicates this balancing act for many water utilities across the United States. Its presence typically leads to the formation of regulated brominated trihalomethanes (THMs) and haloacetic acids (HAAs) in chlorinated waters and forms bromate in ozonated waters.

Bromide is not removed by conventional treatment processes, such as coagulation/flocculation/sedimentation (CFS). Other processes would be required to remove bromide, such as ion-exchange (IX), high-pressure membrane filtration, and electrodialysis. Disadvantages of these processes include the production of a concentrated brine waste stream and high capital and operations and maintenance (O&M) costs. The presence of a brine waste stream can prove very challenging for certain utilities when disposal options are readily available. One emerging technology for bromide removal is electrolysis. Electrolysis of water involves running an electric current through an electrolysis cell. The electrolysis cell consists of two electrodes submerged in an electrolyte (in this case water) and connected to opposite poles of a source of direct current. The oxidation of bromide to bromine occurs at the anode. A complimentary reaction, the oxidation of water at the anode, produces a very low pH around the anode, which creates conditions favoring bromine over hypobromous acid and hypobromite. The oxidation of water also produces fine oxygen bubbles, which leads to bromine volatilization. This overall process leads to decreased bromide levels in the water. Limited research findings and design data on bromide removal by electrolysis are available in the literature. Although critical findings were obtained during the last five years, important data were lacking on electrolytic reactor design and optimization, the potential synergy between electrolysis and conventional treatment processes, and scale-up issues.

PROJECT OBJECTIVES

The goal of this project was to advance the design and understanding of the electrochemical reactor for drinking water treatment applications. The study was focused on bromide and DBP removal, as well as, potential synergies with conventional drinking water treatment processes. Specific objectives were as follows:

- Design a pilot-scale electrolytic reactor based on previous research findings.
- Optimize bromide removal and decrease DBP formation potential.
- Investigate the effect of high power on process efficiency.
- Determine potential synergy between the electrolytic reactor and conventional treatment (i.e., CFS and ozonation).
- Determine effect of electrolytic reactor on microorganism inactivation and on taste and odor (T&O).
- Assess scalability issues of an electrolytic reactor for full-scale drinking water treatment.
- Identify future research needs.

The findings from this project confirm some of the previous findings and assess the feasibility of implementing electrolysis within conventional water treatment plants (WTP).

RESEARCH APPROACH

To accomplish these objectives, the project was conducted in multiple phases.

Phase 1: Reactor Configuration to Optimize Bromide Removal for DBP Control

Lessons learned from previous testing were considered when designing the pilot-scale electrolytic reactor. These include the use of Dimensionally Stable Anode (DSA) for anode material and stainless steel for cathode, perpendicular anode to cathode configuration, the maximization of the anode surface area (i.e., by increasing the number of plates), and allowing the gases formed to naturally strip instead of using air stripping as a second stage.

Phase 2: Reactor Testing

Several experiments were conducted to assess the effect of electric current, hydraulic residence time, and mixing on bromide oxidation, bromine volatilization, instantaneous DBP formation as a result of chlorine/bromine generation, and DBP formation potentials. Other parameters studied included the effect of initial bromide concentration and the effect of dehalogenating the electrolyzed water prior to DBP formation potential testing. Findings from this phase are critical for demonstrability, scaling up issues, and feasibility analysis.

Phase 3: Evaluation of Potential Synergy Between Electrolysis and Conventional Water Treatment Processes

In addition to electrolysis alone, four treatment configurations were mimicked at the bench-scale to evaluate the potential synergies between electrolysis and conventional water treatment processes:

Electrolysis alone
Electrolysis + CFS
Electrolysis + ozonation
Electrolysis + ozonation + CFS
Electrolysis + CFS + intermediate ozonation

For all configurations, electrolyzed waters were subjected to jar testing using variable doses of ferric chloride coagulant with and without ozonation through a bench-scale flow-through ozone contactor. Bromide and bromine were measured in electrolyzed waters, settled waters, and chlorinated waters. Total organic carbon (TOC), pH, alkalinity were measured in settled waters. THMs, HAAs, and total and speciated organic halides (TOX) were measured in chlorinated samples. Bromate was measured in ozonated waters, and T&O characteristics were measured in raw and electrolyzed waters. The findings from this phase of testing help assess whether implementation of an electrolysis process is feasible and beneficial in a conventional WTP.

Phase 4: Scale Up Considerations

The findings from Phases 2 and 3 were used to assess scale up issues. Critical design variables include the required current density (current per surface area of DSA) to achieve a target percent bromide removal or DBP formation potential (DBPFP). A conceptual power cost analysis was conducted. Other safety and full-scale application issues were considered.

Phase 5: Round-Table Workshop to Identify Future Research Needs

A Round-Table workshop was organized towards the completion of the project to determine any outstanding issues and future needs and considerations. These included additional research work and scale-up and safety issues. The expert panel consisted of water quality experts, electrochemists, chemists, engineers, microbiologists, and regulatory representatives.

Phase 6: Mechanistics Effects Testing

Mechanistic effects testing was conducted in parallel to the project to understand the fundamentals behind the electrolysis process, including the irreversibility of the bromide oxidation reaction, the effect of mixing, the effect of bromide concentration, and the effect of pH.

Report Structure

A materials and methods section is presented in Chapter 2. A literature review covering electrolysis principles for bromide removal and findings of mechanistic effects is presented in Chapter 3. The experimental findings, scale up considerations, and future needs are presented in Chapters 4 through 7.

PHASE 1 AND 2 REACTOR CONFIGURATION AND TESTING

A pilot-scale electrolytic reactor (Figure ES.1) was designed to configure a perpendicular anode to cathode mode. Fourteen 3 feet (ft) x 1 ft DSA plates were placed within the reactor. In an initial phase, the reactor was configured to test up to 7 gallons per minute (gpm) of flowrate and up to 35 amps of current. In a later phase, the reactor was tested under high flow/high current conditions (up to 15 gpm of flow and up to 120 amps of current).

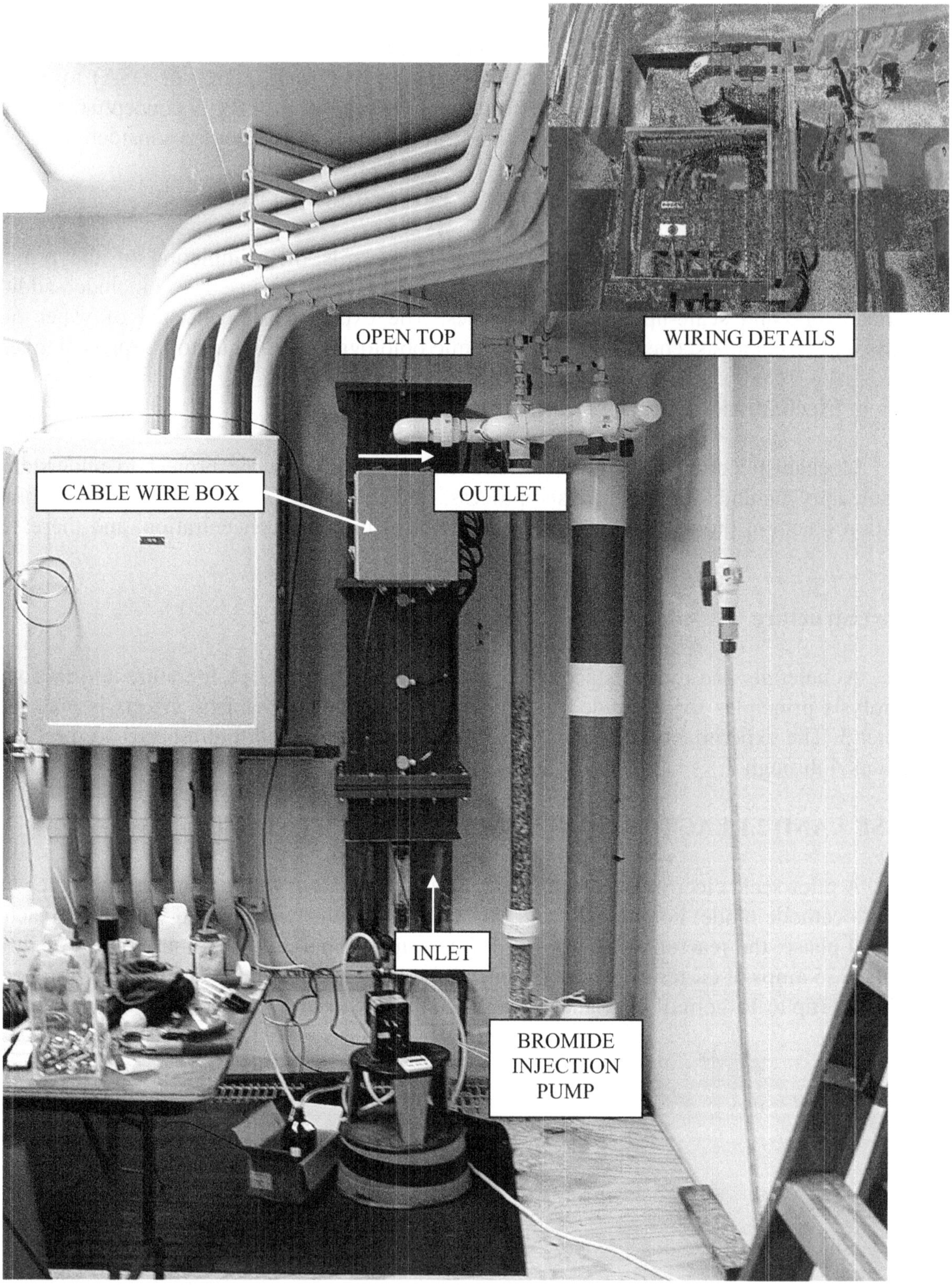

Figure ES.1 Mounted electrolytic reactor during high power testing

Bromide Removal and Bromine Accumulation

Regardless of the initial bromide concentration, bromide oxidation to bromine appeared to be a direct function of the applied current. Bromide oxidation continued as long as bromide was present in the water. On the other hand, the bromine volatilization appeared to be hindered by the remaining bromide concentration in the aqueous phase. As soon the bromide concentration reached a threshold concentration of approximately 50 to 60 µg/L, bromine accumulation was observed. The higher the current applied beyond that threshold, the higher the bromine accumulation.

To achieve high bromide removals (e.g., greater than 60 percent removal), higher current amperage was required with decreasing hydraulic residence time (HRT) and increasing turbulence within the active reactor zone. Without the application of high power, a low HRT and high turbulence combination appeared to significantly hinder the bromide oxidation. As shown by the mechanistics studies and following the principles of Faraday's Law, both HRT and mixing negatively impact the bromide oxidation reaction. Thus, under the higher flowrates tested, sub-optimal conditions for bromide oxidation existed because of low HRTs and the occurrence of mixing.

Optimal current and power ranges were determined under each HRT condition. Optimal conditions were defined as those that resulted in the combined condition of maximum bromide removal with minimal bromine accumulation. The presence of aqueous bromine will lead to the formation of undesirable brominated by-products. These are presented in Table ES.1 and Figure ES.2. Power ranges of 0.2 kW to approximately 2 kW were required for most conditions. The highest flowrate condition which presents high turbulence and low contact time, required 2.4-6 kW of power.

It should be noted that the bromide removals were not equivalent under all HRT and mixing conditions. Even with higher power applied, the bromide removals (µg/L per amp) were significantly higher at the highest tested HRT of 11 minutes (Figure ES.2).

However, limited data have shown that the majority of beneficial bromide oxidation (i.e., volatilization of bromine with limited aqueous bromine accumulation) was actually occurring over the first 10 cm of the DSA plates depth (i.e., equivalent to 1-min HRT), as long as low mixing conditions were present in the reactor. Above that depth, bromine was being accumulated. This could lead to significantly smaller HRTs than shown in Table ES.1, footprint and volume, and DSA surface area requirements.

Despite the fact that sub-optimal bromide removal was measured when 15 gpm current was used (1.5-min HRT), the occurrence of mixing and turbulence within the reaction zone under that higher flowrate may have negatively impacted the bromide oxidation reaction, as discussed in Chapter 2 (mechanistic effects). Thus an HRT of 1 to 1.5 min measured under low turbulence conditions created a better environment for bromide oxidation as opposed to the same HRT under high flow conditions.

Table ES.1
Optimal* current ranges under each tested flowrate condition†

Flowrate (gpm)	HRT (min)	Current (amps)	Power (kW)
2	11.2	15	0.2
4	5.6	15-25	0.2-0.52
5	4.5	<61	>0.52 - <2.6
7	3.2	>35	>0.97 - <2.6
10	2.2	<60	>0.97 - <2.6
15	1.5	58-90	2.4-5.8

*Optimal conditions defined as those leading to maximum bromide removal and minimum aqueous bromine accumulation
†Conditions assuming the full depth of DSA plate

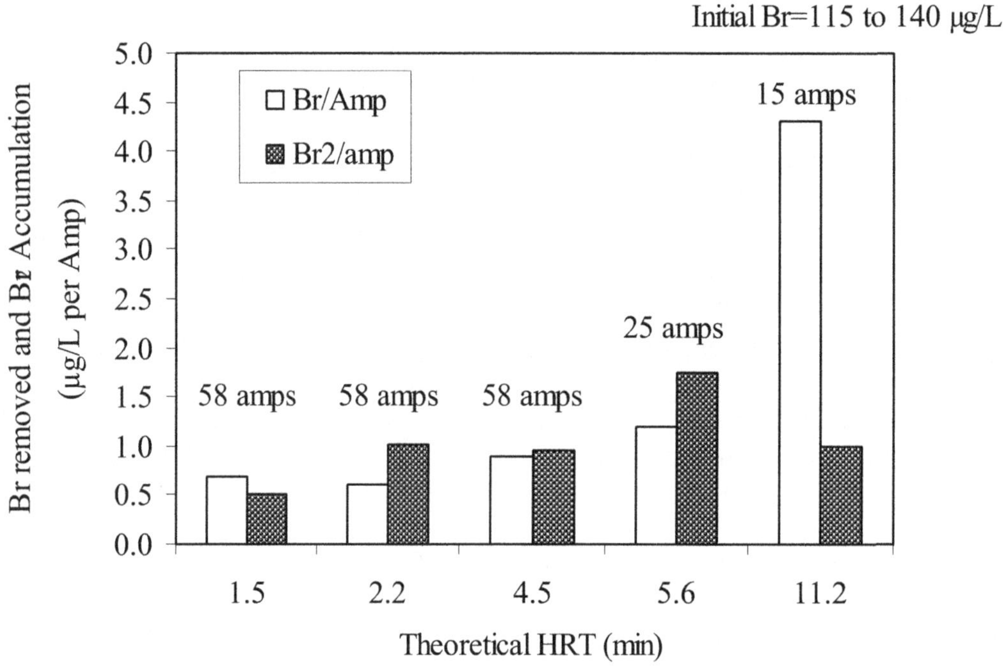

Figure ES.2 Impact of HRT on bromide removal

Instantaneous Formation of DBPs

THMs, HAAs, and TOX are formed during electrolysis. As high as 27 μg/L THM4, 24 μg/L HAA6, and 104 to 141 μg/L TOX concentrations were formed as a direct result of applying 30 to 35 amps of current and under ambient bromide concentrations (150 to 170 μg/L). Free chlorine residuals in those samples ranged from 1.3 to 1.7 mg/L. Under spiked bromide concentrations (i.e., 435 to 470 μg/L), equivalent THM4 and HAA5 levels were formed in the waters electrolyzed at 35 amps, i.e., as high as 30 μg/L THM4 and 26 μg/L HAA6. In addition, similar concentrations of chlorine were formed. Limiting the amount of accumulated bromine/chlorine by applying the optimal current under optimal HRT and mixing conditions is likely to limit the formation of DBPs formed instantaneously.

DBP Formation Potentials

DBPFPs in raw water samples were compared to DBPFPs in electrolyzed water samples, under both ambient and spiked bromide conditions. Similar trends were observed for both conditions. Under ambient bromide conditions (i.e., 140 μg/L), a 10 percent decrease in THM formation potential (THMFP) and TOX formation potential (TOXFP) was observed in electrolyzed samples when 5 to 20 amps of current were applied. Higher applied currents lead to higher DBPFP levels than in raw water. A current amperage of 20 amps resulted in the lowest levels of HAA6FP (12 percent decrease). Under spiked bromide concentration of 435 μg/L, a 30 percent decrease in THMFP was calculated under the optimal current condition of 15 amps. Brominated THM and HAA species were observed to decrease at lower optimal currents (up to 10 amps), and increase thereafter. The bromine incorporation factor (n) in THMs and HAAs, which is the molar ratio of brominated THMs or HAAs divided by THMs or HAAs, was observed to decrease from 1.4 to 0.95 for THMs and from 0.9 to 0.5 for HAAs, when an optimal current of 15 amps was applied. These findings compliment the bromide/bromine data. The higher the bromine accumulation, the higher the DBPFP. It is thus key to limit the amount of accumulated bromine.

To investigate the role of residual bromine in DBP formation potential, selected electrolyzed samples were split and dehalogenated prior to DBP formation potential testing. Dehalogenating the electrolyzed samples decreased the THMFP by 23 μg/L under the same current condition. The drop in THMFP was mainly attributed to a drop in chloroform, dichlorobromomethane, and dibromochloromethane. Thus, limiting the accumulation of aqueous bromine in electrolyzed waters will limit the formation of brominated DBPs.

TOBr and TOCl formation potentials were also measured in raw and electrolyzed waters. For those species, the application of current appeared to decrease the TOBr (by 20 to 50 μg/L), and slightly increase the TOCl (by 4 μg/L to 7 μg/L).

In terms of optimal HRT for DBPFP, the removal of THMFP was approximately five times greater at the highest HRT of 11 minutes and low turbulence conditions as compared to lower HRTs and high turbulence (Figure ES.3).

Bacterial Regrowth Downstream of Electrolysis

Electrolysis results in shifts in the oxidation reduction potential (ORP). It is these shifts in the ORP of the bacteria cells that are thought to suppress the growth of naturally occurring heterotrophic bacteria by reducing the potential and the free energy.

To investigate this hypothesis, raw and electrolyzed samples dehalogenated, sterilized, inoculated with 1 mL of raw water, and incubated over a period of 7 to 21 days. HPCs were used to assess bacterial count. While bacterial growth was observed at the lowest current ranges during the first few days of incubation, growth inhibition was observed in all electrolyzed samples after 18 days of incubation. Significantly lower bacterial counts were measured in samples electrolyzed with currents 25 amps and higher.

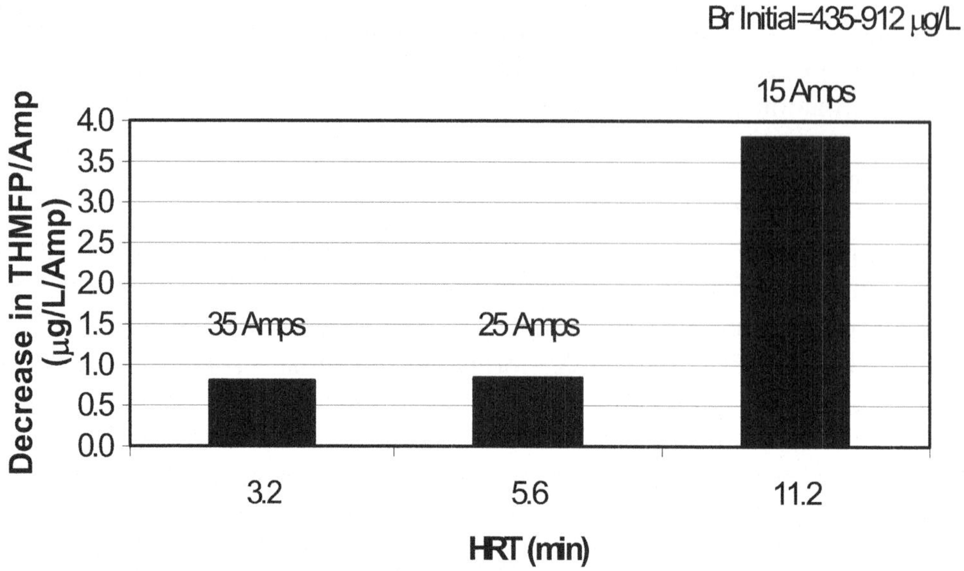

Figure ES.3 Impact of HRT on THMFP removal

While the shift in ORP may play a role in inhibiting bacterial regrowth, more research is needed to identify the exact mechanism of growth inhibition.

PHASE 3: SYNERGY BETWEEN ELECTROLYSIS AND CONVENTIONAL TREATMENT

Four treatment configurations were tested during the course of the study to investigate whether there are any synergistic effects between electrolysis and conventional treatment:

- Electrolysis + CFS
- Electrolysis + Ozonation
- Electrolysis + Ozonation + CFS
- Electrolysis + CFS + Intermediate ozonation

Bromide Removal and Bromine Formation

Some synergistic effects were observed in terms of bromide removal when CFS was applied downstream of electrolysis. The addition of 5- to 20-mg/L ferric chloride to electrolyzed waters increased bromide removals by 14 to 18 percent (bromide decreased from 157 µg/L down to 75 µg/L when 20 mg/L ferric chloride was dosed). However, the fraction of bromine in those samples appeared to increase as well, from 69 percent in the electrolyzed sample to about 90 percent in the CFS-treated electrolyzed sample. These experiments would have to be repeated to check for consistency.

When ozonation and CFS were mimicked downstream of electrolysis, a significant decrease in the fraction of residual bromine was observed (from approximately 70 percent in electrolyzed samples to 10 percent after ozonation+CFS). This may potentially be due to stripping of the aqueous bromine (most likely existing as HOBr under the pH range of 6 to 7) experienced during ozonation and jar mixing conditions, or due to other reactions with ozone to form bromate or total organic bromide. This could in turn lower THM and HAA formation.

Halogenated DBP Formation Potentials

In terms of halogenated DBP formation potentials, CFS is a well-proven treatment process for removing DBP precursors. This was shown in the drop of THMFP in both the CFS-treated un-electrolyzed (from 192 to 36 μg/L) and electrolyzed water samples (from 165 to 61 μg/L). What was interesting however, is the observed synergistic effect of CFS on electrolyzed samples. At 10 amps of current, the THMFPs in the electrolyzed/CFS-treated samples dropped to levels lower than those in the CFS-treated un-electrolyzed samples.

Lower THMFPs (decrease from 192 to 150 μg/L) were measured in the non-electrolyzed sample when 0.08 to 0.3 mg/L ozone residual was mimicked, as opposed to without. The same trend is anticipated in electrolyzed samples. The application of ozone likely oxidized some of the precursors of THM formation, as well as consumed some of the bromide to form bromate rather than brominated THMs.

A decrease in THMFP in electrolyzed samples from 165 to 110 μg/L was measured when ozonation and CFS (at 5 mg/L ferric chloride) were mimicked.

The addition of intermediate ozonation downstream of CFS appeared to somewhat decrease the DBPFP, from 170 to 150 μg/L. The use of electrolysis in a conventional water treatment plant setting appears to be feasible and beneficial.

Bromate Formation

A slight but consistent synergistic effect was observed when ozonation was preceded with electrolysis. A 1.1 to 1.8 μg/L decrease in the bromate formation potential was observed in electrolyzed samples as compared to un-electrolyzed water samples.

The addition of intermediate ozonation downstream of CFS appeared to lower bromate concentrations. Under identical conditions of ozonation and/or CFS, the bromate concentrations in the treated un-electrolyzed waters were 2 to 10 μg/L higher than those in the treated electrolyzed.

Microbial Growth

As presented in Chapter 3, when CFS is added downstream of electrolysis, the same trend in hindered heterotrophic bacteria regrowth is observed in the electrolyzed samples. Regardless of the coagulation conditions, effluent samples electrolyzed at 35 amps measured significantly inhibited growth of heterotrophic bacteria as compared to non-electrolyzed water samples

One potential detrimental effect is the observed increase in the heterotrophic bacteria growth in the ozonated un-electrolyzed and electrolyzed water samples. However, HPC appeared to decrease over time in both non-electrolyzed and electrolyzed ozonated samples.

CONCEPTUAL SCALE-UP CONSIDERATIONS

A conceptual scale-up was conducted involving three types of reactor configurations: Open-channels incorporating DSA plates, DSA ribbons, and DSA wires. The design ratio of required DSA surface area to flowrate varied depending on the HRT, mixing, and current conditions. Assuming the entire DSA depth (i.e., 3 ft) was used, the ratio ranged from 0.05 to 0.38 m^2 per L/min, and the current density on the DSA plates from 0.0005 to 0.0032 amps/cm^2. The limited depth sampling results showed that a 35-amp current and 1/10 of the DSA depth were required under a 2-gpm flowrate condition to obtain bromide removals equivalent to those under the same flowrate, using 15 amps of current and the entire 3 ft of DSA depth. Under this condition, the ratio of DSA surface area to flowrate was 0.04 m^2/Lpm, and the current density was 0.0112 amps/cm^2.

Corresponding full-scale (i.e., 40 mgd) power requirements and required basin volumes were calculated. Because the 2-gpm flowrate and occurrence of minimal mixing were observed as optimal in terms of removing bromide and decreasing DBP formation potentials, the current density and required anode surface area/flow associated with that flowrate were used for design purposes. The design criteria were assessed assuming both the entire DSA plate depth and a fraction (1/10) of the depth used. For both cases, optimizing the configuration of the DSA plates by incorporating DSA ribbons or wires resulted in decreased HRT/footprint requirements. When the analysis was based on the entire depth of DSA plate, even at those lower HRTs (e.g., 3 to 6 minutes) with alternative wire/ribbon configurations, a large footprint would be required. If indeed the majority of beneficial bromide oxidation was occurring over the first 10 cm of the DSA plates depth, the HRT, footprint volume, and DSA surface area requirements could be significantly minimized. The channel length can be decreased at a minimum by an additional 20 percent, and the number of modules is decreased at minimum from 21 to 17. These numbers are based on a 2-gpm flowrate. Higher flowrates would further decrease the channel size.

A direct scale-up to 40 mgd (105,000 L/min) would require power supplies of 3 to 7 Megawatts. Finding a power supply that can output a high current at low voltage is likely to be extremely difficult and would require multiple power sources (as much as 40 units). Research would need to be conducted to determine the availability of power supplies better suited to this task. Power distribution will also require specialized equipment and will likely pose a significant cost.

Safety and maintenance issues will also need to be further investigated. Applying high currents to open channels of water presents significant safety concerns. Effects on downstream processes will need to be identified.

It is, therefore, recommended to conduct further research to better define scaling factors, operational, maintenance, and safety constraints associated with the system. Several configurations of DSAs and several shapes and sizes should be considered to minimize footprint requirements.

PRELIMINARY COST ANALYSIS

System costs were calculated assuming both full DSA depth and a fraction of DSA depth. Electrical equipment is a major contributor of the cost due to the need for specialized power supply, conduits, breakers, and transformers. Assuming a shallow channel design, costs were in the order of \$7.6 million and \$4.6 million assuming the entire DSA depth and a fraction of the depth, respectively. Amortized annual capital costs are estimated between $371,000 and $608,000. Preliminary operating power cost was estimated in $/MG under various ranges of flowrate and for optimal conditions of bromide removal and DBPFP minimization. Power costs were estimated at $200/MG, resulting in $8,000/day to operate a 40-mgd electrolytic system (i.e., $2.9 M annual O&M cost). Total annual costs are estimated at $3.5 M, assuming the full length of DSA plates. When a fraction of the DSA depth is assumed, the power it takes to obtain efficient bromide removal is higher, resulting in higher power cost. As such, operational costs will need to be balanced with the capital costs and overall required footprint.

For comparison, the equipment cost for installing reverse osmosis treatment for bromide removal is estimated at $22 M. This cost does not account for sludge disposal, pretreatment system, a building, or other ancillary features.

FUTURE RESEARCH AND RECOMMENDATIONS

Future research is recommended to confirm the findings from this project, fill in the data gaps, and collect long-term data under conditions more representative of full-scale applications. For other utilities considering the implementation of electrolysis, it will be critical to conduct pilot testing of the process under the ambient water quality conditions and taking into account the existing treatment processes. The limitations or boundaries of the process and ideal conditions for electrolysis will need to be defined on a case-by-case basis.

For the next phase of this specific research, the following aspects should be addressed.

Reactor Configuration and Operation

- Optimal current ranges under various flow conditions need to be better defined. The electrolytic reactor would need to be operated under a narrower range of currents for each flowrate condition. Smaller increments in current/power should also be tested to improve the sensitivity analysis.
- Alternatives to reduce the current requirements in a full-scale system need to be identified, and the reactor design and anode configuration need to be optimized to minimize bromine accumulation.
- Depth samples should be collected to obtain bromide and bromine profiles over the length of the DSA plates. In addition, profile DBPFP data need to be collected over the length of the DSA plates.
- If needed, the reactor may need to be reconfigured with smaller depth DSA plates to truly assess the required current density and ratio of DSA surface area to flow. The flow can then be varied to test the limits of the system.
- One configuration that was not tested is electrolysis downstream of CFS. For those utilities that do not ozonate and have a high concentration of DBP precursors, this configuration could prove beneficial. Under this configuration, lower instantaneous formation of DBPs and DBPFPs would be anticipated.

Water Quality

- Additional DBPFP tests are recommended to confirm the findings from this project. The use of dechlorination should also be repeated to confirm the role of bromine occurrence on DBPFP.
- Speciated TOX (i.e., TOBr and TOCl) formation potentials should be repeated to confirm the findings obtained during the study.
- Additional bromate formation tests should be conducted using a wider range of ozone doses. The optimal current condition for bromate control needs to be assessed.

Costs

- It is recommended to run the electrolytic reactor over a long period to provide data on operational and maintenance costs.
- Identify power consumption, and equipment needs at various voltages.
- Conduct a cost to benefit analysis to assess the feasibility of electrolysis for bromide removal in a full-scale application. What is the minimum HRT and smallest size electrolytic system that can be used to decrease harmful DBPs, while keeping the power costs as low as possible?

Scale-Up Issues

- Evaluate the performance of the electrolytic cell with various anode configurations in order to develop a full-scale design. The shape and surface area of the electrodes and the distance between them could have a significant impact on current density observed at different points along the surface.
- Find a power supply that can output the high current at low voltage required by a full-scale electrolytic cell. Determine if there are power supplies available that are better suited to the task.
- Identify power delivery equipment specialized for this application and identify the costs.
- Define the maintenance needs associated with the operation of the DSA anodes: e.g. intermittent cleaning using chemicals or reverse polarity, periodic removal of modules for cleaning before being returned to service. Estimate the required cleaning methods and intervals for a full-scale system.
- Evaluate how far the current may travel and its potential effects on safety.

To get a better handle on full-scale and safety issues and capital and O&M costs, once repeat tests are conducted at the pilot-scale to confirm some of the findings, it is recommended to scale-up the reactor to demonstration scale (e.g., 3 mgd). At this capacity, more representative capital and operational costs can be obtained. A 3-mgd flow would require one entire module. The number of DSA plates and required surface area in the module would depend on the depth data findings.

CHAPTER 1
INTRODUCTION

PROJECT BACKGROUND

Meeting the disinfectant and disinfection by-product rules has been a challenging balancing act for water utilities as they are required to decrease microbial risk with disinfectants, while simultaneously reduce the chemical risk caused by disinfection by-products. The presence of bromide in drinking water supplies complicates this balancing act for many water utilities across the United States. Sources of bromide include saltwater intrusion, oil-field brines, and industrial and agricultural chemicals (Krasner et al. 1994). Its presence typically leads to the formation of brominated disinfection by-products such as THMs and HAAs in chlorinated waters and limits the use of ozonation for many utilities due to the formation of bromate. With lower maximum contaminant limit (MCLs) for THM4 (80 μg/L) and HAA5 (60 μg/L), as well as, establishing a MCL for bromate (10 μg/L), Stage 1 and 2 D/DBP Rules have resulted in an increasing number of utilities seeking advanced water treatment technologies to reduce DBPs formation, with emphasis on the brominated species.

Key challenges to reducing THMs and HAAs formation during drinking water treatment are the presence of natural organic matter (NOM) and/or bromide. It is widely known that NOM is the principal precursor for these DBPs making it an obvious target for removal during water treatment. Bromide is typically not targeted for removal during drinking water treatment because it is much more difficult to remove, even though brominated species are considered to have greater toxic effects than their chlorinated analogues (Kargalioglu et al. 2000, Zavaleta et al. 1999). In addition, an increase in bromide concentrations has been associated with an increase in THM yield (Chowdhury and Amy 1999). During chlorination, aqueous chlorine (HOCl) oxidizes the bromide to aqueous bromine, which reacts faster and more efficiently with NOM than HOCl (Cowman and Singer 1996, Amy et al. 1991). Moreover, research has shown that the formation of brominated THMs and HAAs, such as dibromochloromethane, bromodichloromethane, and bromodichloroacetic acid, will increase with an increase in the ratio of bromide to TOC or bromide to free available chlorine (Symons et al. 1993, Amy et al. 1991). From this perspective, a reduction in bromide would be a significant advantage in chlorinated systems.

Bromide removal would also be a significant advantage for utilities using or interested in using ozone. Ozone is a widely used technology for microorganism inactivation and taste and odor control, but its use in drinking water treatment is limited by the potential formation of bromate in bromide-containing waters. The implementation of the Long-Term 2 Enhanced Surface Water Treatment Rule (LT2ESWTR), which increases treatment requirements for *Cryptosporidium* based on source water levels, may cause bromide to be a greater problem for utilities using ozone. For example, a utility that is currently using ozone for T&O control and required by the LT2ESWTR to provide *Cryptosporidium* inactivation may not be able to increase the ozone dosage because of bromide levels in the source water. That is to say, any significant increase in the ozone dosage would cause bromate concentrations to exceed the MCL. An economical bromide removal process would allow these previously bromide-bound treatment systems to reap the benefits of ozone disinfection.

TREATMENT TECHNOLOGIES FOR DBP CONTROL

Two main options are available for DBP minimization in drinking water: DBP precursor removal or the use of alternative disinfectants.

Precursor Removal Technologies

Currently best available drinking water treatment technologies for minimizing DBPs (Table 1.1) are often ineffective or costly to build and/or operate and maintain. In the case of bromide removal technologies, there are few choices and those that exist generate large brine waste volumes that require costly disposal methods.

Table 1.1
Brief evaluation of treatment processes to remove DBP precursors

Treatment process	Precursor removed	Advantage	Disadvantage
Enhanced coagulation	NOM	• Low capital cost. • An extension of an established and existing technology.	• Not effective for all waters. • Increased sludge production.
Granular activated carbon (adsorption mode)	NOM	• Can remove precursor to low levels. • Existing facility may be used by retrofitting filters.	• May require high EBCT (> 15 min.) and low surface loading rate (< 3 gpm/ft^2) to achieve adequate NOM removal. • Increase regeneration frequency may be needed to achieve adequate removal. • Carbon replacement or regeneration costs may be high.
Biologically activated media (carbon, anthracite)	NOM	• Inexpensive NOM removal process. • Existing facilities may be used.	• Changes in water quality may disrupt biological activity. • Periodic and/or unexpected spikes in pollutants may disrupt or inactivate biological process. • Removes assimilable organic carbon.

(continued)

Table 1.1 (Continued)

Treatment process	Precursor removed	Advantage	Disadvantage
Ion exchange (including MIEX®)	NOM Bromide	• Can remove NOM and Bromide in same process.	• Produces a waste brine that may be costly to dispose of. • Other water quality parameters may impact removal efficiency. • Concern over NDMA formation may limit selection of resins.
Nanofiltration/ reverse osmosis	NOM Bromide (RO only)	• Can reject multiple organic contaminants. • Removes microorganisms.	• Pretreatment is required. • Produces a large quantity of waste brine that may be costly to dispose of. • High capital and O&M costs.
Electrodialysis reversal	Bromide	• Can reject multiple ionic contaminants.	• Produces a large quantity of waste brine that may be costly to dispose of. • High capital and O&M costs.

Alternative Disinfectants

Several alternatives to free chlorine are available for drinking water treatment, including ultraviolet light (UV), ozone (O_3), chlorine dioxide (ClO_2), and chloramines. All of these alternatives can be effective at minimizing or eliminating DBPs, and present other disinfectant-specific advantages over free chlorine, but each has drawbacks that may make them less practical options.

Ultraviolet Light

UV is typically only cost effective for small treatment plants and provides no residual disinfectant in the distribution system. Reported by-products of UV for disinfection include nitrite (a suspected carcinogen formed from nitrate) (Ijpelaar et al. 2003) and a compound responsible for the destruction of residual chlorine and chloramines in distribution systems, aldehydes including formaldehyde (a suspected carcinogen) (Soroushian et al. 2000), increase in taste and odor (Par et al. 2003), and an increase in mutagenicity of the water (Haider et al. 2003)

Ozone

Ozone provides excellent taste and odor control (Reckhow 1999), some microorganism inactivation, may improve the filterability of colloids, and requires low contact time (Reckhow et al. 1993). However, it provides no residual in the distribution system, and may form bromate in bromide-containing waters (Xie 2004, von Gunten and Hoigne 1994).

Chlorine Dioxide

ClO_2 is a very strong oxidant that requires very low CT values and can provide some T&O control and microorganism inactivation. Chlorine-free ClO_2 does not form THMs or HAAs and can provide a residual disinfectant in the distribution system. However, it is an expensive option that is further limited by the following issues (Xie 2004, Aieta and Berg 1986):

- ClO_2 residual decreases rapidly because of its volatility and reactive nature. Since USEPA regulations mandate utilities to always have a residual in the system, the residual would have to be boosted at several locations (a very expensive practice).
- Possible odor complaints if chlorite is not controlled.
- Chlorite and chlorate formation, requiring additional monitoring.

Chloramines

Monochloramine is the most widely used alternative disinfectant for meeting DBP regulations. Chloramine limits THM4 and HAA5 formation, while maintaining a chlorine residual in the distribution system. However, this approach is not always practical due to the following issues:

- The possible need to feed ammonia at numerous entry points to the distribution system.
- Presence of excess free ammonia from the chloramination process, or from chloramine degradation, can cause nitrification particularly in treated water reservoirs and areas of the distribution system with long residence times. Loss of disinfectant residual, due to nitrification, can lead to proliferation of heterotrophic bacteria, which may include the presence of coliforms (Skadsden 1993, Wilczak et al. 1996).
- Potential formation of DBPs, such as, N-nitrosodimethylamine (NDMA), cyanogen chloride, and cyanogen bromide (Xie 2004, Najm and Trussell 2001).
- Lower inactivation efficiency against microorganisms.

Controlling the challenges presented by alternative disinfectants and currently available DBP precursor removal technologies can be costly to overcome. For some systems, these technologies may simply not be practical or efficient for DBP mitigation.

A promising alternative to these processes is an electrolytic reactor, a selective and cost-effective technology currently used in several industries, including wastewater treatment, metals recovery, and in-situ soil remediation (Bazan and Bisang 2004, Szpyrkowicz et al. 2001, Acar and Alshawabkeh 1993). Electrolytic processes do not require chemicals to operate. In terms of drinking water treatment, an electrolytic process may translate to a low residual producing technology as compared to the technologies listed above with benefits yet unrealized.

TREATMENT CHALLENGES FACING UTILITIES USING STATE WATER PROJECT

California State Water Project (SWP), a source of water for over 23 million people, is typically characterized by low turbidity (2 to 20 NTU), moderate organic matter concentrations

(TOC of 2 to 6 mg/L), and moderate-to-high bromide concentrations (0.2 to 0.4 mg/L). Utilities that rely on SWP as a source of drinking water (such as Castaic Lake Water Agency (CLWA)) often use pre-ozonation, conventional treatment, and secondary disinfection with chlorination. The low turbidity and moderate TOC are usually not amenable to conventional treatment. In addition, operating a sequential system where water is disinfected twice can also pose a challenge in terms of disinfection by-products formation. As a result, very high THMs are formed by the TOC, bromide, seasonally high water temperatures, and long detention times in the distribution system. This contributes to the formation of high levels of DBPs when disinfecting with free chlorine. In 2004, two water systems using finished water produced by CLWA exceeded the 80 μg/L MCL for THM4 set by the Stage 1 D/DBP Rule.

Typically, utilities treating SWP have chosen one of two paths - alternative disinfectants or advanced treatment processes - to minimize DBPs in their water system. Many utilities treating SWP have turned to chloramines as a secondary disinfectant to reduce DBPs in their system in order to meet the Stages 1 and 2 D/DBP Rules. Even though the use of chloramines is an Environmental Protection Agency (EPA)-accepted strategy for minimizing THMs and HAAs, it may present challenges (see above) that make it an unattractive option.

The THM4 and HAA5 MCLs of this rule, implementation of LT2ESWTR, and uncertainties about the use of chloramination will undoubtedly drive an increasing number of water utilities to seek advanced treatment technologies in the future. Some utilities treating SWP have investigated advanced treatment technologies such as IX or granular activated carbon (GAC) processes to meet the D/DBP Rules (Booth et al. 2006). Results have shown that despite the technical feasibility of these processes, regeneration frequencies and residual disposal could make them cost prohibitive when treating SWP to meet the Stage 2 D/DBP Rule. In general, the attractiveness of the advanced treatment technologies may be limited as they are costly, may require the disposal of concentrated residual streams, and/or do not minimize the problems associated with bromide.

Over the past five years, CLWA has been evaluating an electrolytic process as a cost effective approach to remove bromide and minimize disinfection by-products. Kimbrough and Suffet (2002a) have shown that bromide removal with an electrolytic reactor is feasible when treating SWP. In these bench-scale experiments, the reduction in bromide correlated to a significant decrease in THMs, especially the brominated species. Limited pilot-scale testing involved a 3.3-L (Diam: 6.35 cm, height: 2 m) reactor with three 316 stainless steel rod cathodes and four Dimensionally Stable Anodes™ DSA rods (ElTech Systems Corporation, Fairport Harbor, Ohio). In the CLWA studies, a 40-percent reduction in bromide was achieved, leading to THM removal of approximately 60 percent (Kimbrough and Suffet 2002b). Such results show promise for this technology for DBP minimization in drinking water treatment.

The experimental work briefly described above has focused on bromide removal and DBP minimization, showing that an electrolytic process is scientifically feasible for these purposes. However, these studies have indicated that electrolyzed water may also enhance the CFS process without increasing chemical dosages, as well as, provide additional microbial inactivation (Kimbrough et al. 2005, Zhu et al. 2003). Additional microbial inactivation would be a benefit prior to ozone and media filters by reducing the amount of disinfectant or pre-oxidant dosages needed. The potential synergies between an electrolytic process and CFS require further investigation and are incorporated into this proposed project.

Further development of the electrolytic process also requires investigating implementation issues. Experimental work to determine potential synergies with other treatment processes will provide valuable input that may affect process parameters. An examination of the effect of these synergies, as well as, mass and energy transfer changes will help illustrate the viability of an electrolytic process for drinking water treatment.

Overall, an electrolytic reactor may allow water utilities to enhance water treatment without increasing chemical dosages and residual disposal. In other words, it would provide a sustainable technology for drinking water treatment. Previous studies at CLWA have established that an electrolytic process is feasible for removing bromide and minimizing DBPs in drinking water. Further characterization and development (including the determination of scalability issues) of this technology are needed to fully realize its potential. Such development is critical for drinking water utilities, as an electrolytic process may be the key to help water suppliers meet current and future regulations without resorting to costly, waste-producing processes.

PROJECT GOALS

The goal of this project is to advance the design and understanding of the electrochemical reactor during drinking water treatment. The study was focused on bromide and DBP removal, as well as, potential synergies with other drinking water treatment processes. Specific objectives are to:

- Design the reactor to achieve greater bromide and DBP removal than in previous reactor configurations.
- Investigate the effect of high power on process efficiency
- Determine effect of electrolytic reactor on conventional treatment (i.e., CFS and ozonation).
- Determine effect of electrolytic reactor on microorganism inactivation and T&O control.
- Assess scalability issues of an electrolytic reactor for full-scale drinking water treatment.
- Discuss results and identify future research needs.

Addressing the project objectives was conducted in multiple phases:

- Phase 1: Reactor reconfiguration to optimize bromide and DBP removal
- Phase 2: Reactor optimization and testing
- Phase 3: Synergy testing with conventional water treatment processes
- Phase 4: Additional reactor reconfiguration and optimization
- Phase 5: Scale up considerations
- Phase 6: Round-Table workshop to identify future research needs

A materials and methods section is covered in Chapter 2. A literature review of electrolysis principles for bromide removal and of mechanistic work is presented in Chapter 3. The experimental findings, scale up considerations, and future needs are presented in Chapters 4 through 7.

CHAPTER 2
MATERIALS AND METHODS

The following sections describe the materials and methods used to conduct the mechanistic studies, to run the experiments on the electrolytic reactor, to simulate conventional treatment, and to conduct the DBP formation potential tests. Analytical procedures are also described in this chapter.

MECHANISTIC STUDIES

A potentiostat (Model 601A Electrochemical Analyzer, CH Instruments, College Station Texas, USA) was purchased with SCE for a reference electrode, a platinum wire cathode (37 mm long, 0.5 mm in diameter), and a platinum disc anode (1.6 mm wide). This was attached to a personal computer with software that operated the potentiostat and collected and stored the data. Samples were placed into a glass reaction vessel (Figure 2.1) and placed on a small magnetic mixer.

For experiments involving mechanistic studies, a pH 2 Britton-Robinson Buffer (BRB) was prepared and then bromide was added to make a final concentration of 10 mg/L Br- (from NaBr).

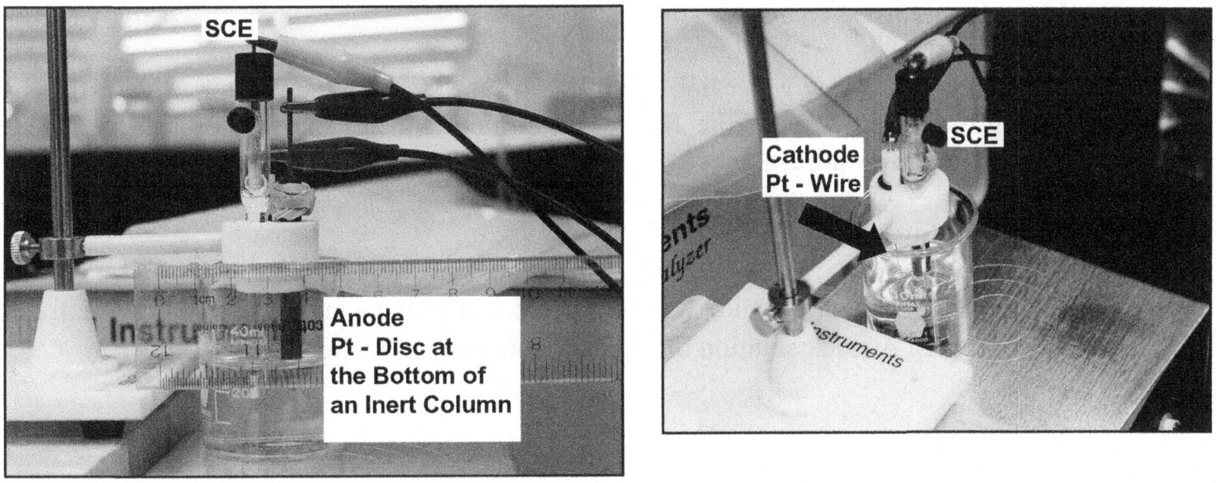

Figure 2.1 Potentiostat & reaction vessel

RAW WATER QUALITY (FEED TO THE REACTOR)

The raw water feeding the electrolytic reactor originated from the SWP aqueduct. Before reaching the two CLWA-owned water treatment plants, the water undergoes a long period of settling in two reservoirs: Pyramid Lake and Castaic Lake, resulting in generally low turbidities. The water quality is summarized in Table 2.1. During the study period, the bromide concentration was well below the historic maximum, leading to the need to spike bromide to the raw water. The target bromide concentration ranged from 300 to 500 μg/L. The TOC concentration was moderate and ranged from 2.5 to 2.7. The specific ultraviolet absorbance (SUVA) values measured during the project period were slightly lower than historic values. SUVA is the ratio of DOC to UV-254 absorbance. A SUVA of 3 is indicative of a high reactivity of the organic matter in the water to form THMs and other DBPs upon chlorination. As a note, based on a raw water TOC concentration and raw water alkalinity, the percent TOC removal under the Stage 1 D/DBP Rule would range between 25 and 35 percent. It should be noted that the organic matter in the water has been proven difficult to remove by conventional treatment plant processes and that a 25 to 35 percent removal would necessitate high doses of coagulant.

ELECTROLYTIC REACTOR

Table 2.2 presents the design features of the electrolytic pilot reactor. A drawing showing the plan and section views of the DSA assembly is illustrated in Figures 2.2 and 2.3, respectively.

Feed into Electrolytic Reactor

A pressurized raw water line (25 PSI) was used to feed the reactor. An existing ball valve provides gross adjustment of the flow. The influent line to the reactor has a globe valve to provide accurate flow adjustment. The raw water feed was fed continuously rather than by batch.

Bromide Spiking Method

In-line spiking of bromide was conducted to achieve a target bromide concentration range of 270-450 μg/L. A stock solution of sodium bromide (1,000 mg/L) was used.

Table 2.1
Historic and current raw water quality

Parameter	Units	Value	Historic average*	Historic median*	Historic 95th percentile*
General chemical and physical					
Temperature	°C		17.8	18.3	21.1
pH	--	7.5	7.9	7.9	8.2
TDS	mg/L		261	254	307
Conductivity	µS	440	511	497	601
Alkalinity	mg/L as $CaCO_3$	80	81	82	90
Total hardness	mg/L as $CaCO_3$		138	137	153
Turbidity	NTU	0.9	0.91	0.71	1.9
TOC	mg/L	2.5 to 2.7	2.5	2.4	3.1
UV-254	1/cm	0.080	0.081	0.077	0.110
Specific UVA (SUVA)	L/m-mg	3.1	3.3	3.2	4.0
Anions					
Chloride	mg/L	45 to 47	59	58	85
Bromide	mg/L	0.12 to 0.17	0.17	0.16	0.27
Nitrate	mg/L	2.3	2.2	2.2	3.2
Sulfate	mg/L	58 to 60	71	63	122
Cations					
Calcium	mg/L	31	28	29	33
Magnesium	mg/L	15	16	15	19
Sodium	mg/L	43	49	52	63
Potassium	mg/L		3.0	2.9	4.0
Detected metals					
Iron	µg/L	51	<20		
Mn	µg/L	4.1	<10		
Background disinfection by-products					
THM4	µg/L	ND	NA	NA	NA
HAA5	µg/L	ND	NA	NA	NA
TOX	µg/L	12	NA	NA	NA

*Data measured between the years 2000-2006
ND: non detected
NA: not available

Table 2.2
Design features of the electrolytic pilot reactor

Overall reactor housing dimensions	1 ft width x 1 ft depth x 4.5 ft height
Flow range	2 to 22 gpm
Theoretical hydraulic retention time	1 to 11 min
Reactor volume from influent line to top	48 gal
Reactor volume from influent to effluent lines	42 gal
Active reaction volume	22.4 gal (3 cu-ft)
DSA plate dimensions (L x W x H)	1 ft x 0.04 in x 3 ft
Maximum number of DSA plates	14
Total surface area of plates	78 sq-ft (7.24 sq-m)
316 SS cathodic plates dimensions (L x H)	1 ft x 3 ft
Minimum space between plates (centerline to centerline)	3 cm—Can be adjusted at 1 cm increments over the 1 ft length
Flexibility in increasing the anode area	Notches at bottom inert support and a matching set of notches on top of DSA plates to allow flexibility in design: Increase the number of plates in the future and easy installation of the DSA plates within the reactor
Minimum space between top of DSA plates and bottom of effluent pipe	2 cm
Flow distribution	Space distributors (three perforated laterals coming from the side) at the bottom of the reactor to prevent short-circuiting between the plates
Flow regulator	Adjustable gate valve on the influent side to regulate the pressurized flow
"Low power" configuration (up to 40 amps)	Two #14 wires with 20-amp capacity each provide current through a fork terminal to the DSA plates (top and bottom, centerlined over the width of the DSA plate). The wires are insulated to avoid dissolution in water.
"High power" configuration (up to 100 amps)	-Addition of two rods -Modification of existing rods -Each rod now has 6 AWG wire on each end for a total of eight -Wires added to the cathode plates for a total of eight -New terminal box provided
Max current density (low flow)	5 Amps/sq-m
Max current density (high flow)	14 Amps/sq-m

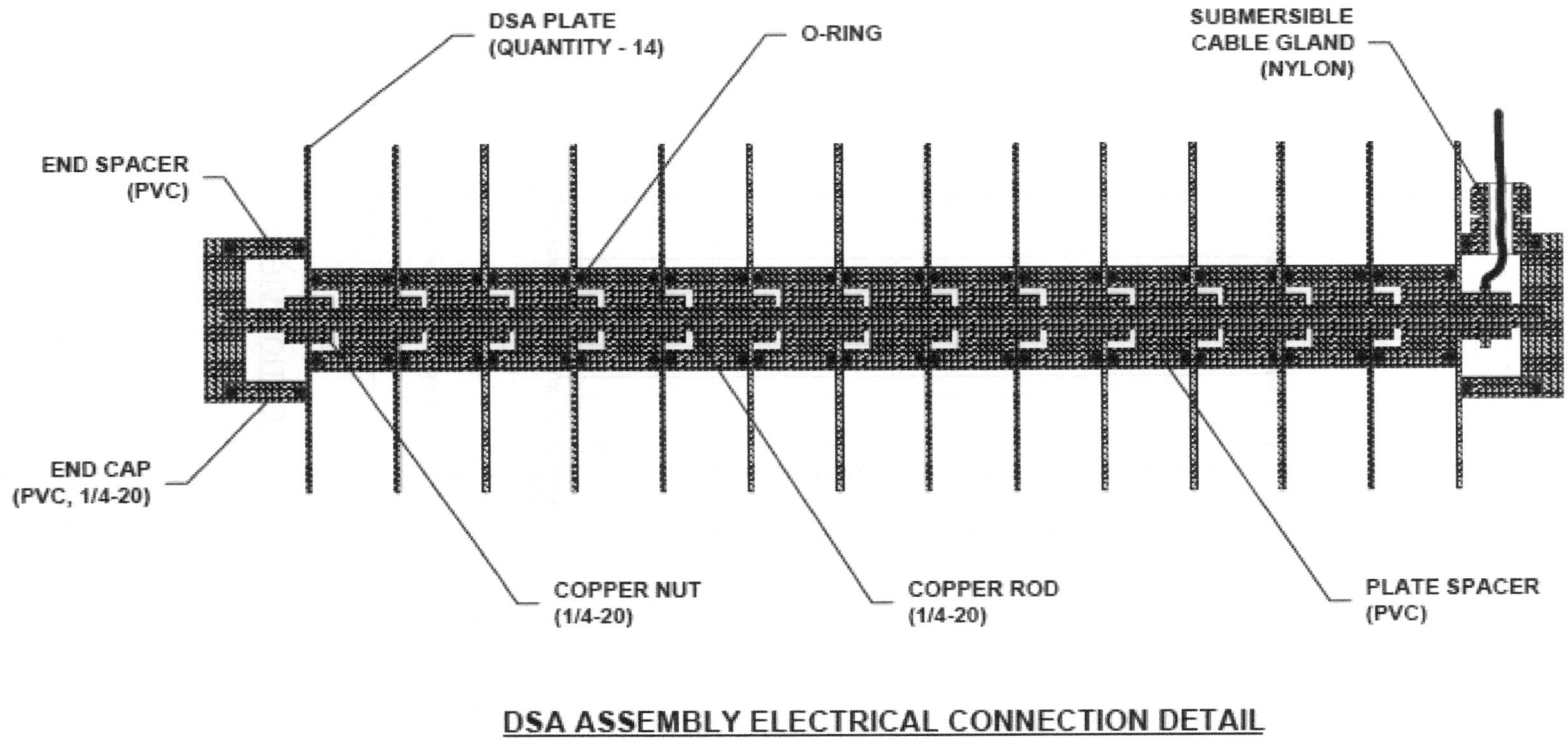

Figure 2.2 Plan view of DSA plate assemblies within the reactor

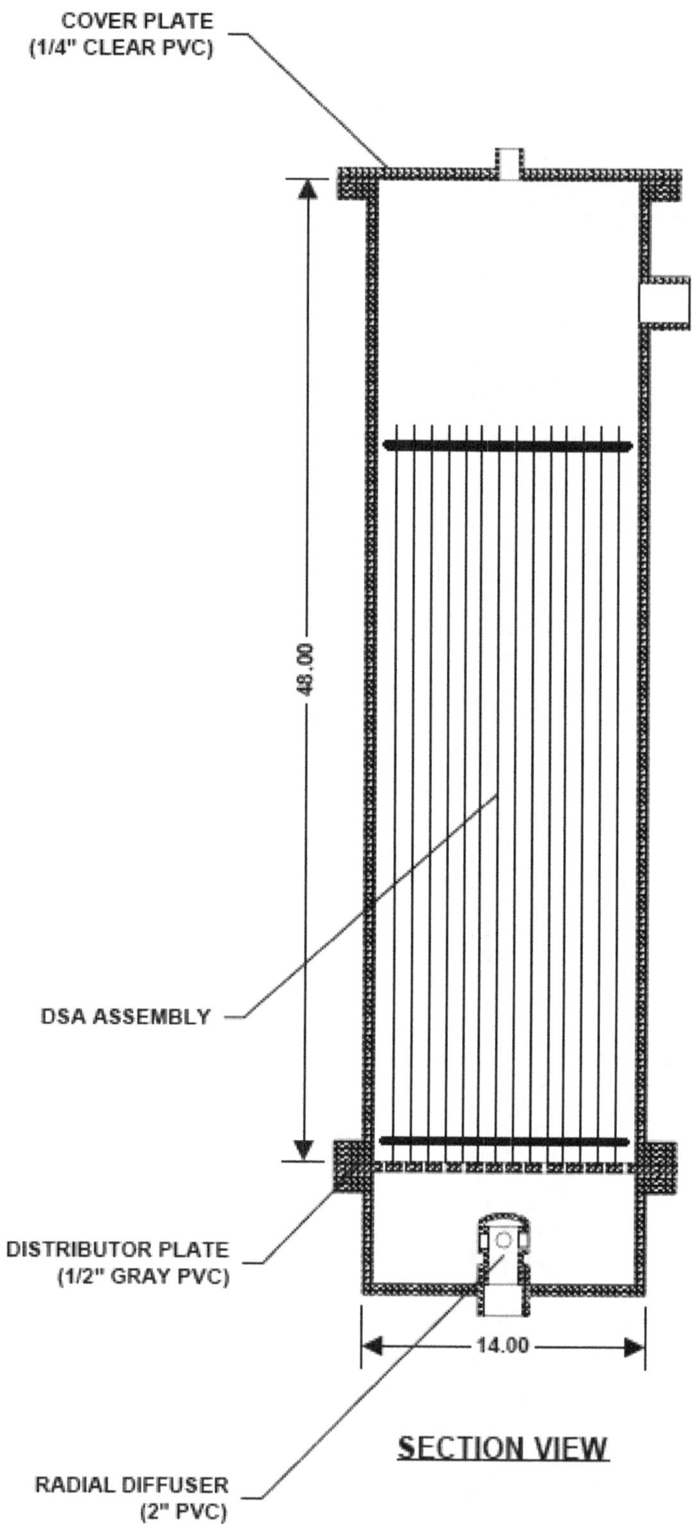

Figure 2.3 Section view of DSA plate assemblies within the reactor

Electrolytic Reactor Conditions During Phase 2

The conditions listed in Table 2.3 were applied in the electrolytic reactor during Phase 2 of testing (electrolytic reactor testing).

Electrolytic Reactor Conditions During Phase 3

The conditions listed in Table 2.4 were applied in the electrolytic reactor during Phase 3 of testing (evaluation of potential synergy between electrolysis and conventional water treatment processes).

SIMULATION OF CONVENTIONAL TREATMENT PROCESSES

The bench-scale experimental design mimicked, as much as practical, the actual conditions applied at the CLWA WTP. Experimental parameters included coagulant type and dose, polymer type and dose, G-values used for coagulation and flocculation, mixing time, sedimentation time, ozone dose and ozonation hydraulic retention time, ozone residual, and chlorination conditions. The existing plant process parameters are summarized in Table 2.5. Rapid mix followed by contact clarification is practiced at the plant.

Since one of the goals of this project was to test conventional CFS post-electrolysis, the jar testing conditions used a more standard design. In addition, to mimic worst-case conditions for DBPFP, free chlorine rather than chloramines was dosed during formation potential tests.

Table 2.3
Experimental conditions applied during low power/high HRT testing

Parameter	Value	Additional comments
Flowrate range	2 to 7 gpm	
Theoretical HRT	3.2 to 11.2 min	
Current amperage	0 to 35 amps	
Power	0 to 950 watts	
Raw water bromide	134 to 150 μg/L	Ambient levels
Raw water bromide	435 to 912 μg/L	Spiked levels after addition of bromide

Table 2.4
Electrolytic reactor conditions

Parameter	Value
HRT	11.2 min
Current	5 to 35 A
Power	30.2 to 945 watts
Bromide concentration	435 to 454 μg/L

Table 2.5
Existing process parameters applied at Rio Vista WTP

Parameter	Value	Notes
Pre-ozonation		
Ozone dose	1.25 mg/L	
Ozone residual	0.3 mg/L	Measured after first chamber
Ozonation HRT	1.5 min at max flow 7.5 min at min flow of 7 mgd	Measured in first three cells
Ozonation T/To ratio	0.50	
Conventional treatment		
Coagulant and dose	Ferric chloride, 1.3 mg/L	This dose is based on the low turbidity in the raw water
Polymers and doses	• Cationic polymer, 1.5 mg/L as product (takes into account 40% activity) • Cationic polymer, 0.25 mg/L	• Flocculation aid added upstream of clarifier • Filtration aid
Coagulation G value	900 sec^{-1}	
Coagulation mix time	1 sec	
Flocculation number of stages	Not applicable	
Flocculation G value	Not applicable	
Flocculation mix time	Not applicable	
Sedimentation time	Not applicable	Upflow contact clarification is practiced
Contact clarifier detention time	9 min	Surface loading 8.3 gpm/sq-ft
Chlorination		
Total chlorine dose	2.5 mg/L	Free chlorine added post filter Ammonia added to clearwell influent
Residence time	24 h	
pH	8.1	Caustic added after ammonia addition

Jar Testing Conditions

Conventional treatment consisting of CFS was undertaken at the bench-scale using a Phipps and Bird standard apparatus utilizing 2.8-L square jars, an impeller drive, and impellers. In this project, the conditions summarized in Table 2.6 were used and held constant throughout the project period.

Continuous Flow Bench-Scale Ozone Reactor

A bench-scale ozone unit was used, consisting of three chambers ozonating in counter-current flow for ozone transfer, mimicking the performance of a variety of full-scale ozonation contactors (Figure 2.4). The system combined an ozone-air mixture with the water provided. One or two contactors were used. The contact time was adjusted by changing the flowrate, and the ozone concentration varied depending on the output of the generator.

Table 2.6
Jar testing conditions

Parameter	Value
Ferric chloride doses	Low dose testing: 0.5, 1.5, and 5 mg/L High dose testing: 5, 10, and 20 mg/L
Cationic polymer dose	1.5 mg/L
Rapid mix RPM	300 RPM
Rapid mix time	20 sec
Four stage tapered flocculation RPM	80 RPM; 60 RPM; 40 RPM; 20 RPM
Flocculation time per stage	5 min each phase
Settling time	20 min (equivalent to 0.125 gpm/sq-ft)
Filtration of settled water samples prior to FP chlorination tests	Through 10-µm filter paper

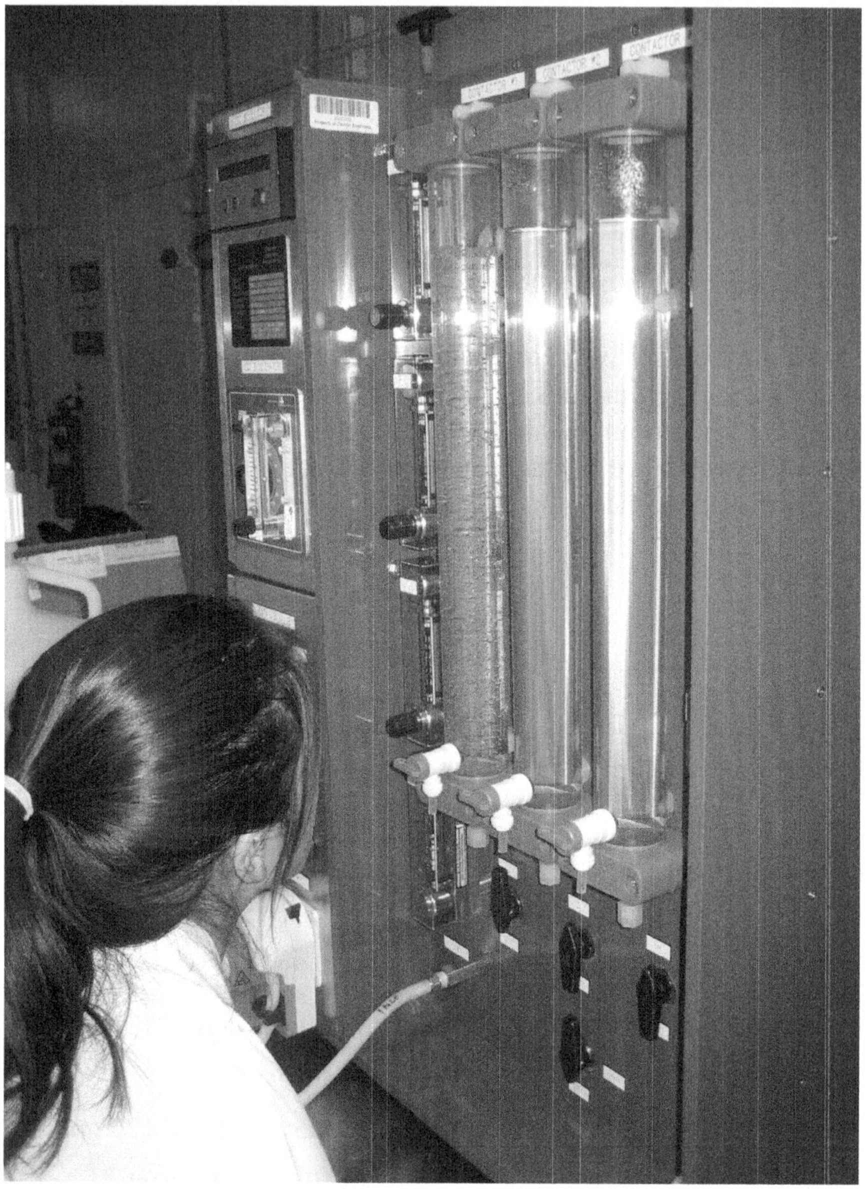

Figure 2.4 Continuous flow ozone bench-scale reactor

Chemical Reagents

Several chemical reagents were employed during the course of this study. These are summarized in Table 2.7.

Table 2.7
Chemical reagents employed in the study

Reagent	Details	Application and Vendor
Ferric Chloride	SG 1.34% as $FeCl_3$	Coagulation From Rio Vista WTP
Sodium Hypochlorite	10-14% by volume	Chlorination
Sulfuric Acid 10N solution		pH adjustment and sample preservation
Sodium Hydroxide 10N solution		pH Adjustment
Cationic Polymer as product	SG 1.03	Flocculation and filtration aid From Rio Vista WTP
Bromide as sodium bromide	1000 ppm or dry powder	Spex, NJ

ANALYTICAL PROCEDURES USED DURING REACTOR TESTING AND CONVENTIONAL TREATMENT

All analytical measurements conducted in this study were based on standard or EPA approved-methods (Table 2.8), with the exception of the specialized measurements discussed in the following sections.

Prior to collecting any samples from the reactor effluent, an equivalent of three hydraulic retention times was allowed when initiating a reactor run, or when a current or flow condition is changed.

Table 2.8
General analytical methods used in the project

Parameter	Analytical Method	MRL	Laboratory
pH	EPA 150.1	0.01 pH	CLWA
Turbidity	EPA 180.1	0.05 NTU	CLWA
Temperature	SM 2550	0.5°C	CLWA
Alkalinity	SM 2320B	2 mg/L	CLWA
Hardness	EPA 200.7 by calculation		CLWA
TDS	SM 2540C		CLWA
Conductivity	SM 2510B		CLWA
ORP	SM 2580		CLWA
Metals	EPA 200.7		CLWA
HPC	SM 9215		CLWA

(Continued)

Table 2.8 (Continued)

Parameter	Analytical Method	MRL	Laboratory
TOC	EPA 415.1	0.5 mg/L	CLWA
Filtered UV-254	SM 5910B	0.009 1/cm	CLWA-Spectrophotometer
Bromide	EPA 300.0B	5 μg/L	CLWA
Chloride	EPA 300.0A		CLWA
Free Bromine	EPA 300.0B[a]		CLWA
Sulfate	EPA 300.0A		CLWA
Total Cl_2	SM 4500 CL	0.2 mg/L	Hach Kit
THM4	EPA 524.2	0.5 μg/L	CLWA
HAA5	6251B	1-4 μg/L	MWH Lab
TOX	EPA 9020	1-4 μg/L	MWH Lab
TOBr and TOCl	Method developed as part as a AwwaRF project (Hua and Reckhow, 2005)		David Reckhow Laboratory at U. Massachusetts
O_3	SM 4500-O3	0.1 mg/L	Spectrophotometer
T&O	SM 6040D	1 ng/L	UCLA (Mel Suffet)
Bromate	EPA 317 (low level)	1 μg/L	MWH Lab

* Measured following addition of sodium thiosulfate to reduce bromine to bromide.

Free Chlorine

Free chlorine residual was measured using a Hach pocket colorimeter and the DPD colorimetric method (based on Standard Methods 4500-Cl DPD Colorimetric Method). It should be noted that this method does not differentiate between bromine and chlorine. However, the bromine concentration is in orders of magnitude lower than the chlorine concentration (μg/L for bromine versus mg/L for chlorine).

Free Available Bromine

Duplicate samples were collected. Bromide was measured in the first sample by ion-chromatography, and an excess of sodium thiosulfate ($Na_2S_2O_3$) was added to the second sample. $Na_2S_2O_3$ reduces any bromine, hypobromous acid, or hypobromite that is present to bromide. Bromide was measured in the second sample. The difference between the bromide in the sample quenched with $Na_2S_2O_3$ and the unquenched sample is the free available bromine concentration. This method is detailed in Kimbrough and Suffet (2006).

Ambient Chlorine/Bromine Gas

Ambient chlorine and bromine gas concentrations in the headspace over the DSA plates were monitored using a Draeger Tube Kit with a range of 0.2-3 mg/L of chlorine. It should be noted that no gas was detected within the concentration range in any of the experiments.

Instantaneous DBP Formation

DBPs formed in samples as a direct effect of electrolysis were measured on-site in all electrolyzed samples (reactor effluent) without quenching of the chlorine/bromine residual.

Disinfection By-Products Formation Potential Tests

DBPFP tests were conducted by applying a free chlorine dose to the samples to achieve a 10-mg/L final chlorine residual at time zero and incubating the samples for 24 hours at room temperature. For electrolyzed samples, which already contained a free chlorine residual, chlorine was added to achieve the 10 mg/L target. Settled water samples were vacuum filtered through a 10-µm filter paper prior to chlorination.

After the 24-hr incubation period, the samples were transferred to THM, HAA, and TOX bottles containing the appropriate reducing agent and the chlorine residual was` measured. It should be noted that in these tests, the chlorine formed in the sample as a result of electrolysis was not quenched prior to DBPFP testing.

A second series of DBPFP tests were later conducted to evaluate the role of bromine and chlorine in the formation potential of DBPs. In these series of tests, electrolyzed samples were split in halves, with only one of the halves receiving a dose of reducing agent to quench the chlorine/bromine formed as a result of electrolysis. Both samples were then subjected to DBPFP.

Halogenated Total Organic Halide (TOBr and TOCl)

Samples were sent to the University of Massachusetts for TOBr and TOCl analyses. The methodology was published in Hua and Rechow (2005).

Flavor Profile Analysis

Select settled water samples were collected and sent to the University of California, Los Angeles (Dr. Mel Suffet) for flavor profile analysis (Standard Method 2170).

Batch Ozonation Experiments

One series of batch ozonation experiments was conducted in which electrolyzed water samples were quenched with first hydrogen peroxide, then shipped to Arizona State University for ozonation testing. Prior to electrolysis, bromide was spiked to three levels: 220, 270, and 400 µg/L. Spiked raw water samples were electrolyzed at currents of 0, 15, 25, and 35 amps under a 11.2-min HRT. Once on site, samples were ozonated to achieve three target residuals: 0.5, 1.5, and 3 mg/L. Samples were allowed to react for a period of 4 days.

CHAPTER 3
ELECTROLYSIS PRINCIPLES AND MECHANISTIC EFFECTS WITH EMPHASIS ON BROMIDE REMOVAL

ELECTROLYTIC PROCESS

An electrolytic process consists of two electrodes (anode and cathode) and an electrolyte. An electrical current enters at the anode and exits at the cathode. The electrolyte is typically a liquid that completes the electrical circuit between these two electrodes. The electrons that are transferred between the electrodes and an electrolyte will produce electrochemical oxidation (Equation 3.1) and reduction (Equation 3.2) reactions at the anode and cathode surfaces, respectively.

$$A \rightarrow B + ne^- \quad (3.1)$$

$$B + ne^- \rightarrow A \quad (3.2)$$

The electrode potential drives these reactions, and the applied current will determine the rate of reaction. An increase in the applied current generally results in a higher rate of reaction. The relationship between the applied current (or current density) and a chemical species, *i*, can be described by Faraday's Law:

$$m_i = It/nF \quad (3.3)$$

where m_i = the molar flux of chemical species *i* (mol/cm^2)
I = current density (A/cm^2)
t = time (seconds)
n = stoichiometric ratio of electrons for conversion of species *i* (moles)
F = Faraday's constant (96,487 coulombs/mole)

The electrode potential, E_{App}, needed to achieve the applied current density is defined as:

$$E_{App} = E_{MT} + E_{Kin} + E_{Sol} \quad (3.4)$$

Where E_{MT}, E_{Kin}, and E_{Sol} are the potentials associated with mass-transfer effects, the electrode kinetics, and solution resistance, respectively. In general, these three components of the applied potential should be minimized for an electrochemical system. E_{MT} is affected by the movement of a chemical species between the surface of the electrode and the bulk solution and can be reduced with an increase in turbulence, or mixing, in a reactor. E_{Kin} is a function of the electrode material and can be minimized by choosing a material appropriate for the application (i.e., water, wastewater, metals, etc.). E_{Sol} is a function of the water conductivity and can be reduced by increasing the specific conductivity of the water, increasing the electrode surface area in the reactor, and/or decrease the distance between the electrodes.

The use of an electrolytic reactor is a novel approach to drinking water treatment for bromide removal and DBP minimization. Limited research has been conducted on the electrolytic removal of bromide, as well as electrolytic enhancement of coagulation and microbial inactivation. The following subsections summarize the available research literature on these topics.

Electrolytic Bromide Removal

Kimbrough and Suffet (2002a) determined the feasibility of using an electrolytic reactor for removal of low concentrations of bromide and the minimization of trihalomethanes when treating SWP. They identified the half-cell reaction for bromide oxidation at the anode as:

$$2Br^- \rightarrow Br_2 + e^- \quad E^o = -1.1 \text{ V} \tag{3.5}$$

Once formed, if bromine is not volatilized, it will be hydrolyzed and can exist in one of the three states, molecular bromine (Br_2), hypobromous acid (HOBr), or as hypobromite ion (OBr-).

$$Br_2\,(g) \leftrightarrow Br_2 + H_2O \leftrightarrow HOBr + H^+ \leftrightarrow OBr^- + 2\,H^+ \tag{3.6}$$

The equilibrium constant for hydrolysis of molecular bromine at 25°C has been measured between 5.2 and 7.2 × 10^{-9} M^{-2} at pH 7 (pKa = 8.7). If the pH is greater than 3.4, (HOBr is the dominant state and OBr^- is dominant when the pH is greater than 9. If the pH is less than 3.4, bromine volatilization takes place. It is thus critical to maintain a low pH around the anodes. This is achieved with the parallel reaction of oxidation of water at the anode:

$$2H_2O \rightarrow O_2 + 4H_+ + 4e^- \quad E^o = -1.23 \tag{3.7}$$

Another important oxidation reaction that takes place at the anode is the oxidation of chloride into chlorine:

$$2Cl^- \rightarrow Cl_2 + 2e^- \quad E^o = +1.36 \text{ V} \tag{3.8}$$

As shown in Equation 3.8 chlorine is generated during the process of electrolysis of raw water containing chloride, or during the electrolysis of brine solutions.

In Kimbrough and Suffet's study (2002a), bench-scale experiments were conducted with a 550 mL (5-cm inside diameter), undivided, up-flow electrolytic reactor containing three carbon rod cathodes (diameter = 6.25 mm) and four DSAs. The DSAs are titanium rods (6.25-mm diameter) coated with a proprietary mix of TiO_2 and RuO_2. Experiment 1 demonstrated stable electrode voltages after the equivalent of three hydraulic residence times.

One experiment showed the effect of electrolysis on the bromide speciation when treating SWP. Experiments were performed over a range of constant applied currents (0 to 800 mA). Effluent bromide was observed in the forms of bromide ion, hypobromous acid/hypobromite, and brominated THMs (bromodichloromethane, dibromochloromethane, and bromoform). Results showed that as the applied current increased from 0 to 800 mA, bromide ion concentrations decreased from 200 μg/L to non-detect, hypobromous acid/hypobromite concentrations increased from non detect to 180 μg/L, and brominated THMs increased from non detect to 10 μg/L. At each applied current, the total mass of bromide in the effluent was within 15 percent of the influent mass. The authors hypothesized that the remaining 15 percent of bromide mass was volatilized as bromine gas (Br_2) as a result of low pH, then carried out of the reactor by oxygen bubbles formed at the anodes.

In another experiment, SWP was electrolyzed over a range of applied currents (0 to 800 mA) and subsequently sparged with carbon dioxide to pH 5. The THM formation potential of the resulting solutions were evaluated at pH 8.4, 24-hour contact time, 20°C, and 5 mg/L free Cl_2 (added). Results showed that as the applied current increased from 0 to 800 mA, the THM_4 formation potential decreased from 210 to 180 μg/L and the incorporated bromide concentration decreased from 102 to 72 μg/L.

Based on these results, the authors believed the electrolytic removal of bromide during drinking water treatment to be a promising approach to minimized DBPs, especially the brominated species.

Kimbrough and Suffet (2002b) continued the work described above and began to explore the practicality of using an electrolytic reactor for bromide removal and THM minimization when treating SWP. Six pilot-scale experiments were conducted with a 3,300 mL (6.35-cm inside diameter), undivided, up-flow electrolytic reactor containing three 316 stainless steel rod (6.25-mm diameter) cathodes and four DSAs. The DSA were the same diameter and composition as described above. Effluent from the electrolytic reactor entered the top of an 8-m high air stripper (15.2-cm diameter). Air and carbon dioxide were mixed with a gas proportioner and fed to the bottom of the air stripper with a high volume diffuser.

The first experiments examined the challenges and solutions to comparing DBP formation potential in un-electrolyzed and electrolyzed water. These experiments were conducted with only the electrolytic reactor and DBP formation potential tests as needed. Electrolyzed water was produced at flowrates of 0.75 to 3 L/min with applied currents of 4 to 7 amps. The authors concluded the following:

- Due to the formation of hypobromous and hypochlorous acids during electrolysis, the standard methods for conducting DBP formation potential were not adequate to compare un-electrolyzed and electrolyzed waters.
- To more accurately compare electrolyzed water with un-electrolyzed water, a reducing agent, such as sodium thiosulfate, could be added to electrolyzed waters in order to reduce hypobromous and hypochlorous acids prior to a DBP formation potential tests.
- "Self-stripping" of bromine gas appears to occur in the electrolytic reactor.

Subsequent experiments examined post-electrolysis volatilization of bromine gas using the air stripper. Electrolyzed water was generated at flowrates of 1.5 to 20 L/min with applied currents of 0 to 35 amps. Carbon dioxide was injected in the air stripper at rates of 0 to 1080 L/min. From these experiments, the authors showed the following:

- The self-stripping of bromine gas observed in the previous set of experiments was reproduced at each flowrate. The amount of self-stripping was proportional to the contact time.

- The current needed to completely oxidize bromide increases with decreasing reactor contact time.
- Up to 40 percent of the influent bromide could be removed from SWP with electrolysis followed by stripping with carbon dioxide.
- There was no apparent correlation between gas flow in the stripping process and removal of bromide.

Coagulation Enhancement

No research was identified in peer-reviewed journals showing the enhancement of coagulation by electrolysis during drinking water treatment. However, the following studies suggest that coagulation enhancement using electrolysis is achievable via electro-aggregation or electropolymerization of dissolved organics.

Yang et al. (2004) used a bench-scale, undivided electrochemical reactor with graphite, stainless steel, and active carbon fiber (ACF) electrodes to remove low-molecular weight fulvic acid by electro-aggregation from a model water. Initial experiments showed that a maximum removal of fulvic acid in the model water at an electrode potential of 20 V and each subsequent experiment was conducted at this potential. Results from the model water study showed that the ACF electrodes were the most effective for electro-aggregation of fulvic acid. Particle diameters began increasing from < 10 nm after 30 minutes of contact time to 459 nm after 240 minutes of contact time. During electrolysis, the fulvic acid concentration decreased slightly but stayed relatively constant at 80 mg/L as carbon. After 240 minutes of contact time, $Fe(OH)_2$ was added to the electrolyzed water (with no applied electrode potential). This solution was adjusted to pH 8 and allowed to settle for 30 minutes. The TOC decreased to 13.5 mg/L. Based on these results, the authors proposed a mechanism for fulvic acid removal using electrolysis with ACF:

1. Fulvic acid adsorbs to the ACF surface.
2. Larger aggregates form electrochemically.
3. The larger aggregates desorb from the ACF surface.
4. The larger aggregates are coagulated with $Fe(OH)_2$ and removed.

Further testing with a fulvic acid solution derived from Hui River (China) sediment showed 100 percent removal of fulvic acid after 60 minutes of electrolysis with ACF electrodes.

Matsushita et al. (2005) studied the electropolymerization of low concentrations of aniline (5.0×10^{-5} M) in an aqueous, phosphate buffered solution (0.1 M) at neutral pH. The bench-scale, flow-through electrolytic cell consisted of two compartments, one for the anode and the other for the cathode, separated by a Vycor glass tube. Water (with aniline and phosphate) flows through the anode compartment while the cathode compartment is filled with phosphate buffer solution. Platinum, glassy carbon, and carbon fiber anodes were tested for their ability to electropolymerize aniline. Both carbon-based anodes were shown to electropolymerize aniline on their surface at electrode potentials ranging from -0.2 to 1.15 V, unlike the platinum anode, which showed no polymerization of aniline. Subsequent testing with the high surface area carbon fiber (3700 cm^2 of fiber/cm^3 of reaction volume) showed that up to 90 percent of the aniline was removed from the solution over the course of one hour of treatment at an electrode potential of 0.8 V.

Enhancing Microorganism Inactivation

Bacterial metabolism and cellular integrity are maintained by balancing the redox state of all the cellular components for optimal overall function (Green and Paget 2004). However, disturbance of this balance, notably an increase in oxidation reactions (oxidative stress), can damage essential cellular components. Consequently, bacteria have evolved sensitive and specific sensors to monitor different redox signals (Green and Paget 2004). Electrolysis results in shifts in the redox potential of microorganisms without the addition of chemicals. The ORP shift is enough to suppress the growth of naturally occurring HPC bacteria. Carlson (1991) reviewed 51 papers on disinfection and reported several where the ORP had a significant impact on disinfection dynamics (Table 3.1). The ORP can be changed without the addition of oxidizing or reducing agents by electrolysis. Limited research was identified showing enhanced microbial inactivation using electrolysis. However, the following studies suggest that increased microbial inactivation can be achieved with electrolysis during drinking water treatment.

Zhu et al. (2003) compared the virus and NOM removal efficiency of conventional coagulation and electrocoagulation pretreatment of a microfiltration membrane. A model and Lake Houston waters were treated at the bench scale with ferric iron (chemically added or electrochemically generated) as the coagulant followed by a hydrophilic, flat-sheet microfiltration membrane (0.22-μm pore size, 4.1 cm^2 surface area) operated in dead-end filtration mode. Results showed that electrocoagulation consistently outperformed conventional coagulation for virus and NOM removal efficiency for ferric iron concentrations of 2-10 mg/L and pH 6.3-8.3. Virus removals as high as 4.9 log and 4.4 log were observed for electrocoagulation and conventional coagulation, respectively, at 10 mg/L Fe^{3+}. The maximum NOM removal observed was 41 percent for electrocoagulation with about 8 mg/L Fe^{3+} and 36 percent for conventional coagulation with 10 mg/L Fe^{3+}. The authors hypothesized that the superior performance of electrocoagulation is due to the attraction of negatively charged viruses and NOM to the positively charged anode.

Kimbrough et al. (2005) studied the effect of changes in ORP on bacterial growth. A bench-scale, undivided electrolytic cell (6.1-cm diameter, 30.5-cm height) with three graphite cathodes (6.25-mm rod diameter) and four dimensionally stable anodes (6.25-mm rod diameter) was used to change the ORP as a function of applied current in a model water (sodium sulfate solution with a conductivity of 1000 μS, halogen free). Heterotrophic plate counts (HPC) were used to assess the bactericidal effects of electrolysis after 1 mL of raw water from Castaic Lake (Calif.) was added to the solution. Results showed that the change in ORP of the model water is directly proportional to the electrolysis time. Furthermore, a linear relationship was observed between HPC directly after inoculation and the ORP of the electrolyzed water. As ORP decreased from 300 to -300 mV, HPCs decreased from approximately 3,000 to 55 CFU/mL. A second-order relationship was observed between ORP (300 to -300 mV) and the regrowth of bacteria as measured by HPC after a 24-hour incubation period at 20°C (approximately 30,000 to 180 CFU/mL, respectively). The authors concluded that the shift in ORP of the electrolyzed water may have disrupted the metabolic pathways of the bacteria and that the greater the shift in the ORP of the water from the optimal ORP for bacterial growth, the greater the suppression of that growth.

Table 3.1
Effect of ORP on contact time to achieve 3 logs kill of *E. Coli*

Range of ORP (mV)*	Contact Time (min)
+450-500	167
+500-550	6
+550-600	1.7
+600-650	0.5

Table adapted from Carlson 1991
*Relative to saturated calomel electrode (SCE)- ORP was changed by adding HOCl

MECHANISTIC EFFECTS OF BROMIDE OXIDATION BY ELECTROLYSIS

In order to better understand the mechanistic effects behind the electrolytic process, both a literature review and a series of experiments were conducted as in-kind and parallel to this project.

Literature Review of Mechanistic Effects

Six papers were reviewed to investigate the mechanistic effects of bromide oxidation and reduction (Table 3.2). The complete list of authors is provided in the list of References. The oxidation of bromide occurs at approximately 1.0 V vs. Saturated Calomel Electrode (SCE), and reduction at 0.8 V vs SCE. The reaction is quasi reversible. When the bromide ion contribution to current reaches the applied current, the potential shifts to that of the next available species (H_2O) to maintain the current. Both reactions (i.e., bromide oxidation and H_2O oxidation) take place until all bromide is oxidized. After that, only H_2O is oxidized. H_2O oxidation is critical to get the low pH required to favor the bromine volatilization. Therefore, using the minimal amount of current for bromide oxidation and for H_2O oxidation is the key.

Table 3.2
Summary of bromide electrochemistry: Potentials for oxidation and reduction vs SCE*

Ref†	Dryhurst	Nematollahi	Allen	Lowe	Lowe	Lowe	Lowe	Zang	Wu	Wu
Electrode	Pyrolytic	GCE	Pt vs Ag	Edge	Basal	GCE	BDD	C-nano-chitosan	BDD	GCE
Br^- to Br_2	0.97V	0.98V	1.1V	0.99V	1.15V	1.03V	1.74 V			
Br_2 to Br^-	0.76-0.87V	0.65V	1.5V	0.83V	0.56V	0.78V	Irrev.	0.71V	0.56V	0.79V

* Saturated Calomel Electrode, unless noted otherwise
†Complete list of authors provided in the list of references at the end of the report
GCE: glassy carbon electrode
BDD: boron doped diamond
Pt: platinum
Ag: silver

Mechanistic Experiments

The following paragraphs describe the findings of the mechanistic experiments. The key results are summarized in Table 3.3.

Table 3.3
Summary of key findings of mechanistic effects testing

Experiment no.	Key finding
Experiment 1a	Under a constant bromide concentration of 10 mg/L and pH of 2, the electrical current peak during bromide oxidation moved towards somewhat more negative potentials (0.90 to 0.825 volts vs. SCE) with increasing scan rates (volts/sec), indicating an irreversible reaction.
Experiment 1b	Higher electrical current peaks were obtained using the platinum anode as opposed to a glassy carbon anode.
Experiment 1c	The current is most intense at pH 2 and decreasing toward pH 6, with pH 8 and 10 being about the same.
Experiment 1c	The oxidation of bromide to bromine appears to occur most readily at pH of less or equal to 2. The oxidation of water to oxygen and hydrogen ions, which occurs during bulk electrolysis, would drastically lower the pH, facilitating the oxidation of bromide.
Experiment 2	For both mixed and unmixed samples of BRB, the current was at the maximum at the initial application of potential and declined rapidly. The mixed samples showed lower currents than the corresponding unmixed samples except for the two lowest concentration samples, 0.1 and 0.2 mg Br^-/L, where they were equal.
Experiment 3	Increased stirring reduced the change in current slightly and reduced the maximum current measured. Summing the area under the curve (amps x seconds or coulombs/ second x seconds) shows considerably larger number of coulombs consumed in the oxidation of bromide to bromine in the unstirred sample than the stirred samples. The results suggest that quiescent waters improves the oxidation rates of bromide to bromine in a Britton-Robinson pH 2 buffer as compared to stirred water. This effect may be due to the disruption of the electrical double layer between the Pt anode and the buffered water.

Experiment 1 Effect of Scan Rate, Electrode Material, pH, and Bromide Concentration

During Experiment 1, several mechanistic aspects were investigated, namely: a) effect of scan rate, b) effect of electrode material, c) effect of pH, and d) effect of bromide concentration.

a) Linear Sweep Voltametry (LSV) experiments were conducted using a narrow range of voltages (1.5 to $^{-}$1.5 volts). A BRB was prepared and bromide was added to make a final solution with a concentration of 10 mg/L Br^- solution (pH 2). This solution was then electrolyzed at different scan rates from 0.025 to 1.00 Volts/second (V/s) and sensitivity of 0.000001 (Amps/Volts). The results are shown in Figure 3.1. The peak height increased with increasing scan rate until about 10 V/s, which was very close to maximum. The peak moved towards somewhat more negative potentials (0.90 to 0.825 volts vs. SCE) with increasing scan rates, indicating an irreversible reaction.

b) This same experiment was conducted using Cyclical Sweep Voltametry (CSV) using both a glassy carbon disc anode and the platinum disc anode at three different scan rates (0.05, 0.10, and 0.50 V/sec). A peak was observed on the glass carbon anode at 0.7 volts vs. SCE, while on the platinum anode it was observed at 0.9 volts vs. SCE. The peaks appeared somewhat stronger on the platinum anode as compared to the glassy carbon, suggesting better conditions for bromide oxidation when a platinum anode is used.

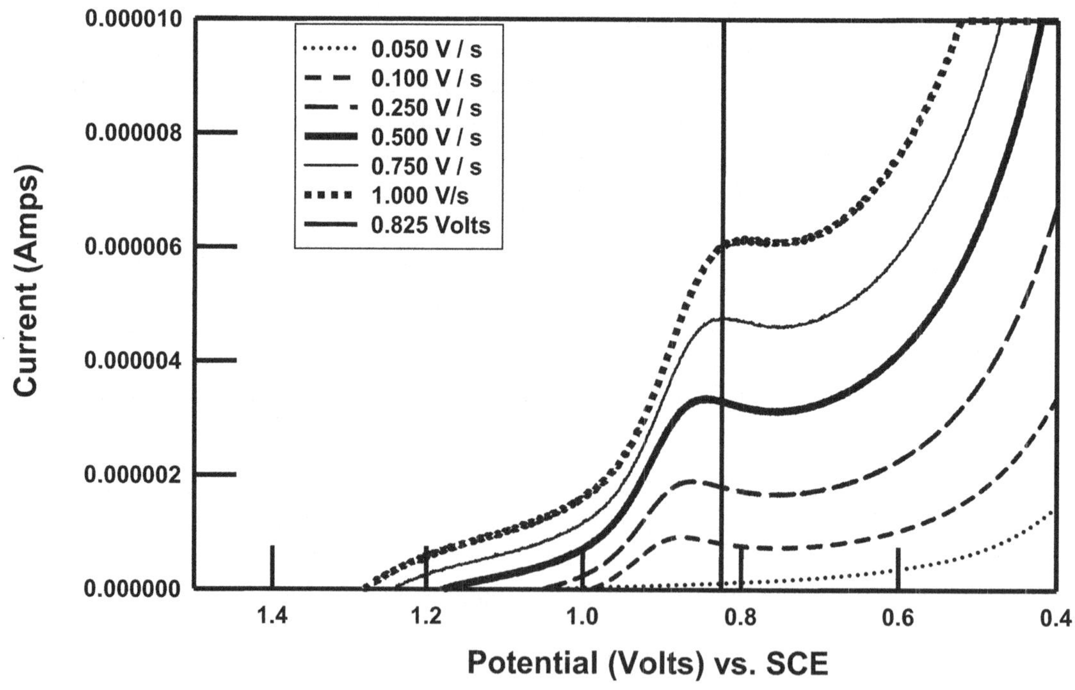

Figure 3.1 Effect of scan rate on current during bromide oxidation

c) The same solution was then divided up into several aliquots and each had its pH adjusted with NaOH with a range from 2 to 10. Each was analyzed on a potentiostat at a rate of 1.0 V/s and sensitivity of 0.00001 (Amps/Volts). In Figure 3.2, the current was most intense at pH 2 and decreased toward pH 6, with pH 8 and 10 being about the same. This confirms that a low pH environment is important for bromide oxidation.

The peak heights at 0.825 volts (vs. SCE) were plotted against the pH in Figure 3.3. The oxidation of bromide to bromine appears to occur most readily at pH of 2. The oxidation of water to oxygen and hydrogen ions, which would occur during bulk electrolysis, would drastically lower the pH, facilitating the oxidation of bromide.

d) A similar buffer was made up at pH 2 and then spiked several aliquots with 1, 2, 5, 7, and 10 mg/L Br- and ran under the same conditions as above (Figure 3.4). As expected, the current peak height was a function of concentration.

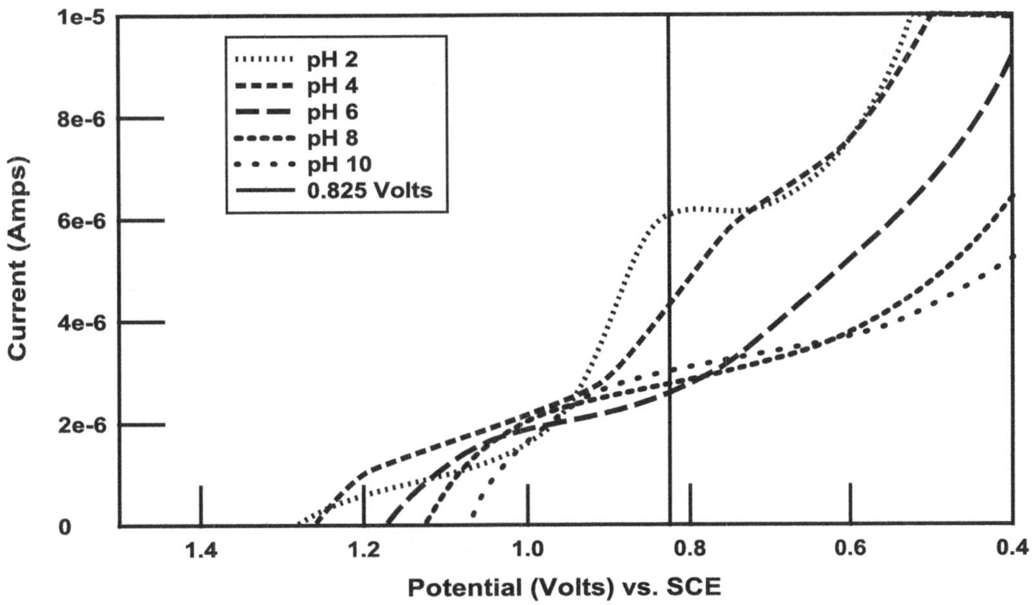

Figure 3.2 Effect of pH on current during bromide oxidation

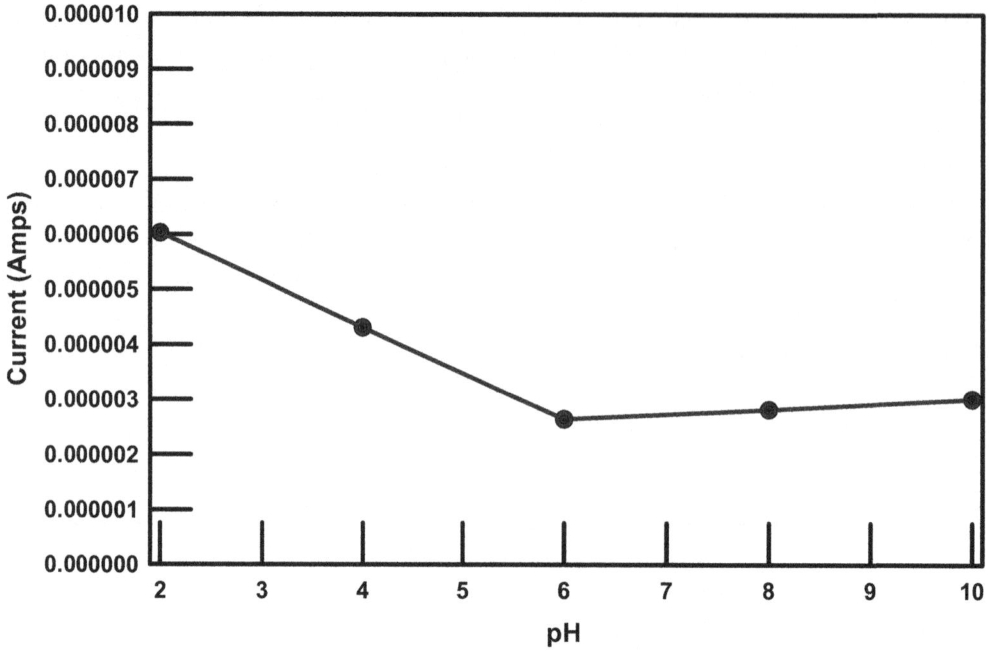

Figure 3.3 Current peak height at 0.825 V versus pH during bromide oxidation

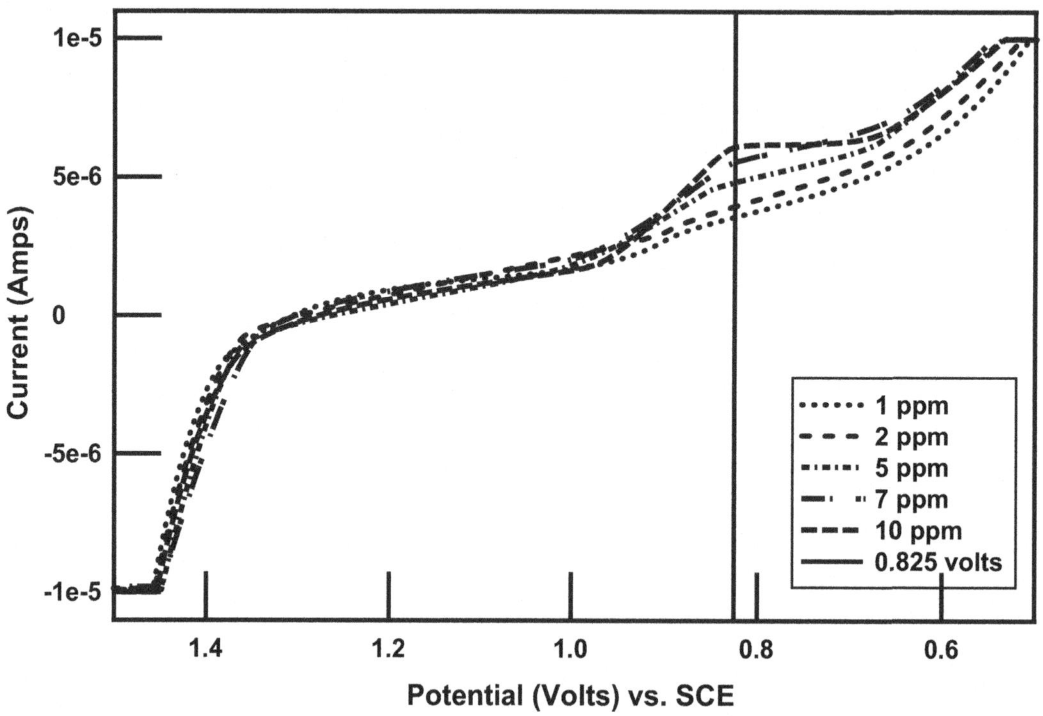

Figure 3.4 Effect of bromide concentration on current during bromide oxidation

Experiment 2 Effect of Mixing of Bromide Oxidation

In Experiment 2, the effect of mixing was investigated under variable concentrations of bromide and under constant potential.

The potential was held constant at -0.8 volts vs. SCE using the platinum electrode and the change in current was measured over time. This was done at several concentrations with and without mixing. The same Britton-Robinson buffer (pH 2) was prepared at several concentrations, 0, 0.1, 0.2, 0.5, 1.0, 2.0, 5.0, and 10 mg Br^-/L. This was then placed in the reaction vessel and the potential was applied. The current was read with the magnetic mixing bar turned off, and the experiment was repeated with the magnetic stirring rod on. In both cases, the current was at the maximum at the initial application of potential and declined rapidly (Figure 3.5). The mixed samples showed lower currents than the corresponding unmixed samples except for the two lowest concentration samples, 0.1 and 0.2 mg Br^-/L where they were equal. The current as a function of concentration is shown at two different sample points, 0.5 and 1.2 seconds. These results would suggest that quiescent waters improves the oxidation rates of bromide to bromine in a Britton-Robinson pH 2 buffer as compared to stirred water. This effect may be due to the disruption of the electrical double layer between the Pt anode and the buffered water.

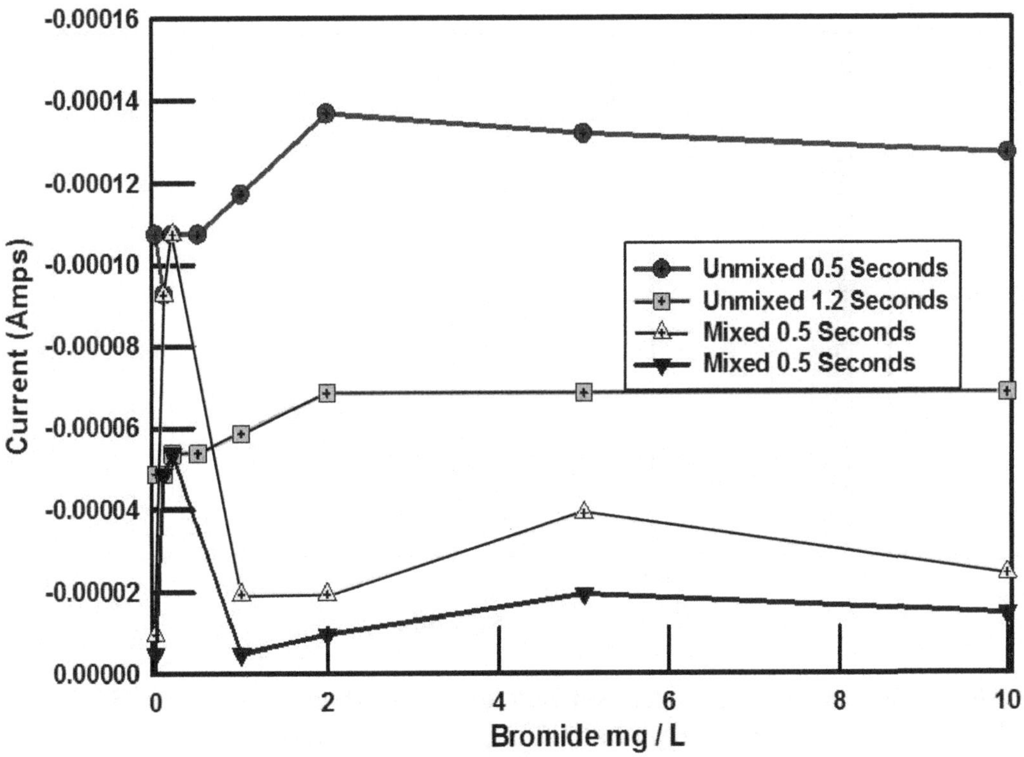

Figure 3.5 Effect of mixing on current at two times, at different bromide concentrations, and at constant potential of –0.8 volts versus SCE

Experiment 3 Effect of Mixing Speeds on Bromide Oxidation

In Experiment 3, the same BRB (pH 2) was prepared at 1.2 mg Br/L. 9 mL of this solution was then placed in the reaction vessel and the potential was applied and the current was read for 2 seconds with the magnetic mixing bar turned off. This experiment was repeated with the magnetic stirring rod on at different stirrer settings (1, 2, 3, and 4). The exact rpm for each setting is unknown, but they were visibly different. The results are shown in Figure 3.6.

As can be seen, the maximum current change (at 0.02 seconds) was reduced in the stirred samples versus the unstirred sample. Increased stirring reduced the change in current slightly. This same relationship was observed at all times less than 0.5 seconds. In Figure 3.7, the current vs. stirring setting is shown for two time periods, 0.02 and 0.20 seconds. Stirring greatly reduced the maximum current measured. Similarly, the stirred samples dropped below 0.00001 amps in less than 0.5 seconds while the unstirred sample had a current higher than 0.00001 amps for the entire period of the experiment (2 seconds). Summing the area under the curve (amps x seconds or coulombs/second x seconds) show considerably larger number of coulombs consumed in the oxidation of bromide to bromine in the unstirred sample than the stirred samples.

These results would again suggest that quiescent waters improves the oxidation rates of bromide to bromine in a Britton-Robinson pH 2 buffer as compared to stirred water.

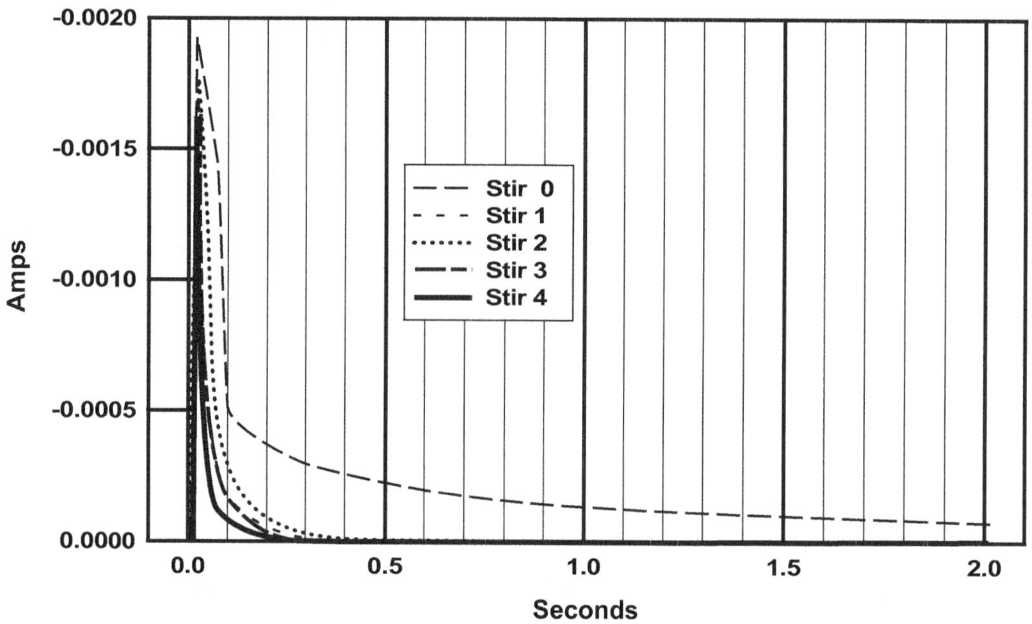

Figure 3.6 Effect of mixing speed settling on current changes over time at constant potential of –0.8 volts versus SCE and at 1.2 mg Br/L in a BRB

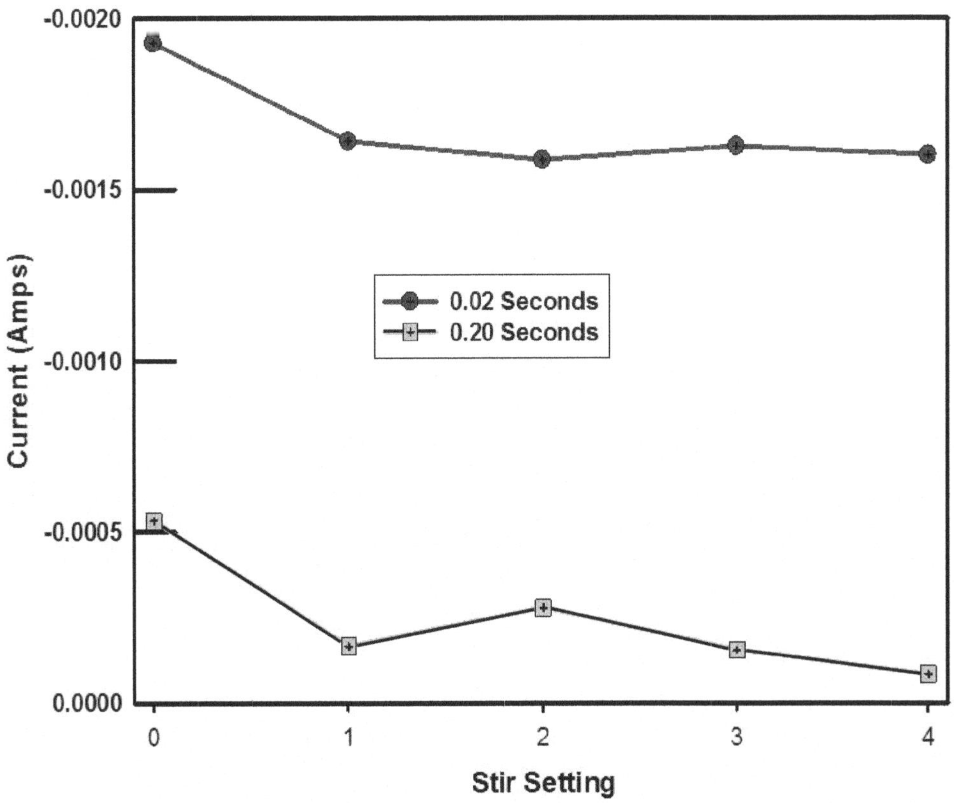

Figure 3.7 Effect of mixing speed setting on current changes measured at two times, at constant potential of –0.8 volts versus SCE and at 1.2 mg Br/L in a BRB

CHAPTER 4
REACTOR DESIGN AND TESTING

ELECTROLYTIC REACTOR DESIGN

Basis of Electrolytic Reactor Design

Lessons learned from previous work conducted by Kimbrough and Suffet (2006, 2005, 2002 (a) and (b)) were applied during the configuration of the pilot-scale electrolytic reactor. In past research, several prototypes of the electrolytic reactor were tested at the bench-scale. The following factors were taken into account:

- Use a perpendicular anode and cathode configuration to minimize the distance between the anode and cathode and reduce the resistance for economic feasibility.
- Increase the anode surface area by placing several flat plates over the length of the reactor (similar to lamella or tubular clarifiers concept): this increases the areas of low pH (less than 3.4), which promotes the formation of bromine gas versus aqueous hypobromous acid (HOBr) or hypobromite (OBr^-).
- Previous research by Kimbrough and Suffet (2002 (a) and (b)) has shown that the distance between the plates makes no difference on the reaction rates on a $/cm^2$ basis. The number of plates is however critical (as discussed in the point above).
- Mechanical mixing is not required since mixing is provided by the fine oxygen bubbles and bromine gas formed at the anode from the oxidation of water and bromide, respectively.
- No air-stripper is required since air stripping of bromine gas occurs naturally as long as the pH in the vicinity of the DSA plates is kept less than 3.0.

Since the depth of the reactor was not determined as a factor affecting reactor efficiency, the reactor depth was kept small. The anode and cathode materials used in previous experiments by Kimbrough and Suffet was used in this project: DSA coated anode plates and stainless steel cathode plates.

During the course of reactor optimization and testing, the reactor was re-wired to accommodate for up to 100 amps to achieve high bromide removals/DBP potential reduction, especially at higher flowrates. Conceptual profile and plan view drawings of the re-designed reactor were developed. Photos of the reactor are shown in Figure 4.1 and Figure 4.2.

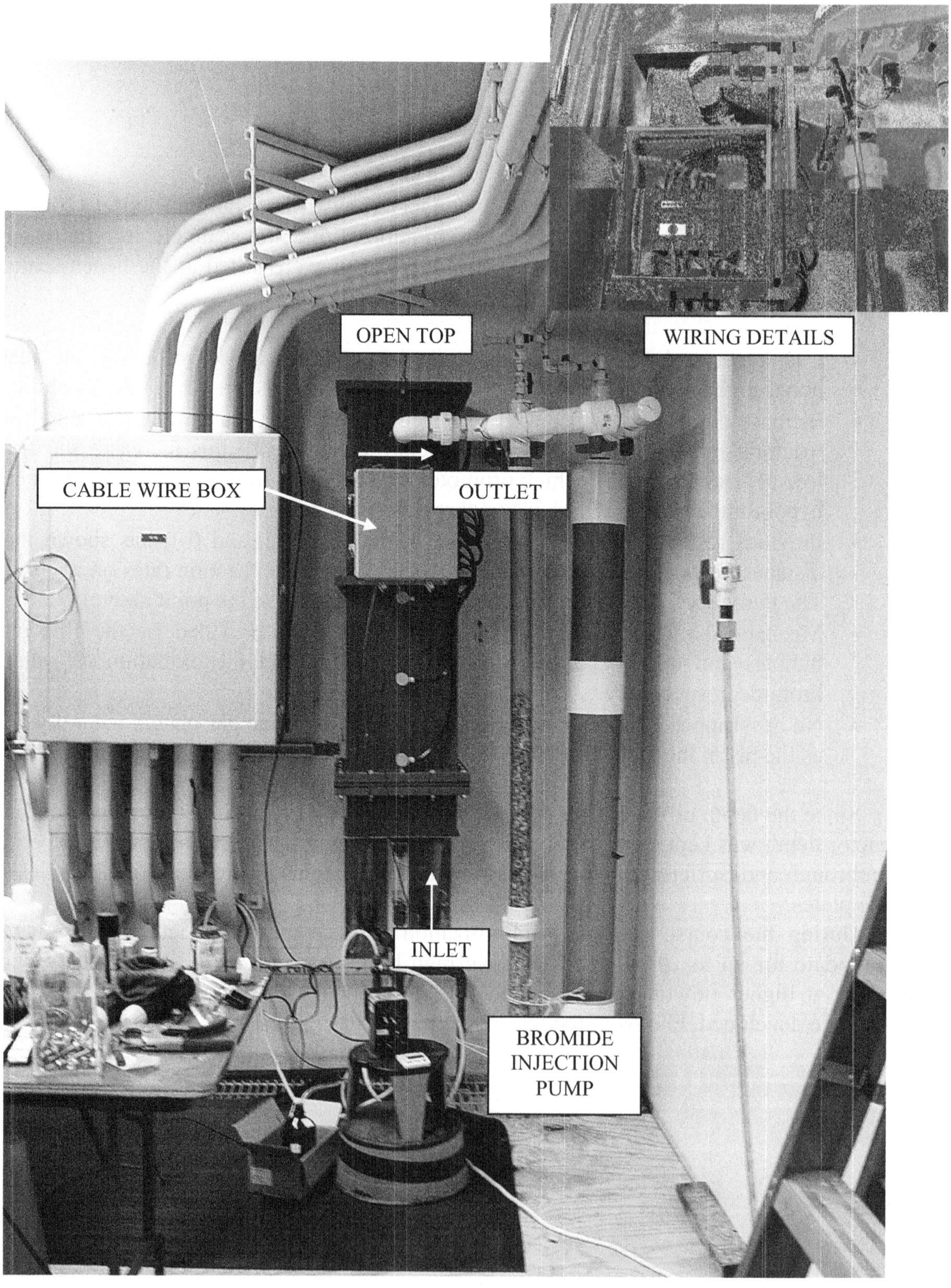

Figure 4.1 Mounted electrolytic reactor during high power testing

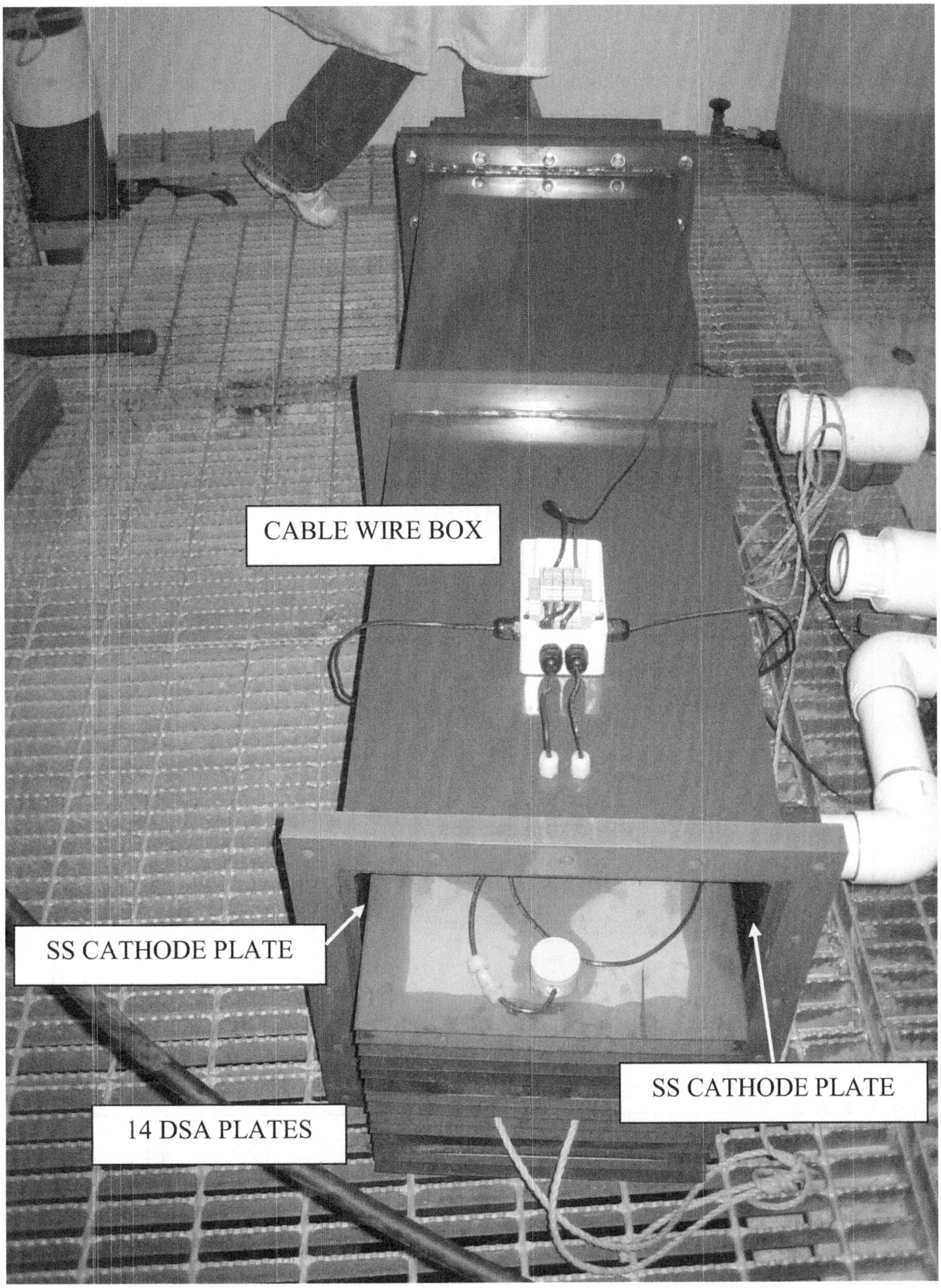

Figure 4.2 Configuration and wiring of DSA plates during low power testing

Tracer Test

A step-dose tracer test was conducted to verify the T_{10}/T ratio and the hydraulic conditions in the reactor. The test was conducted by continuously injecting bromide into the reactor, and measuring bromide samples every 30 seconds over a period of 14.5 minutes. The tracer (bromide) concentration was adjusted for the background concentration (136 μg/L). The flowrate was set at 10 gpm, and the theoretical detention time was calculated at 4.2 min. The tracer curve is plotted in Figure 4.3. The C represents the measured concentration and the Co is the bromide dose. The T_{10}, which occurs when C/Co=0.1 was estimated at 2.2 min, resulting in a T_{10}/T ratio of 0.52. This ratio is indicative of average baffling hydraulic conditions and existence of some short-circuiting within the reactor.

REACTOR TESTING FOR BROMIDE REMOVAL

A series of electrolysis tests was conducted under low flow/low power condition and later under high flow/high power condition to assess the efficiency of the process and the range of operations for bromide removal and abatement of DBPFP.

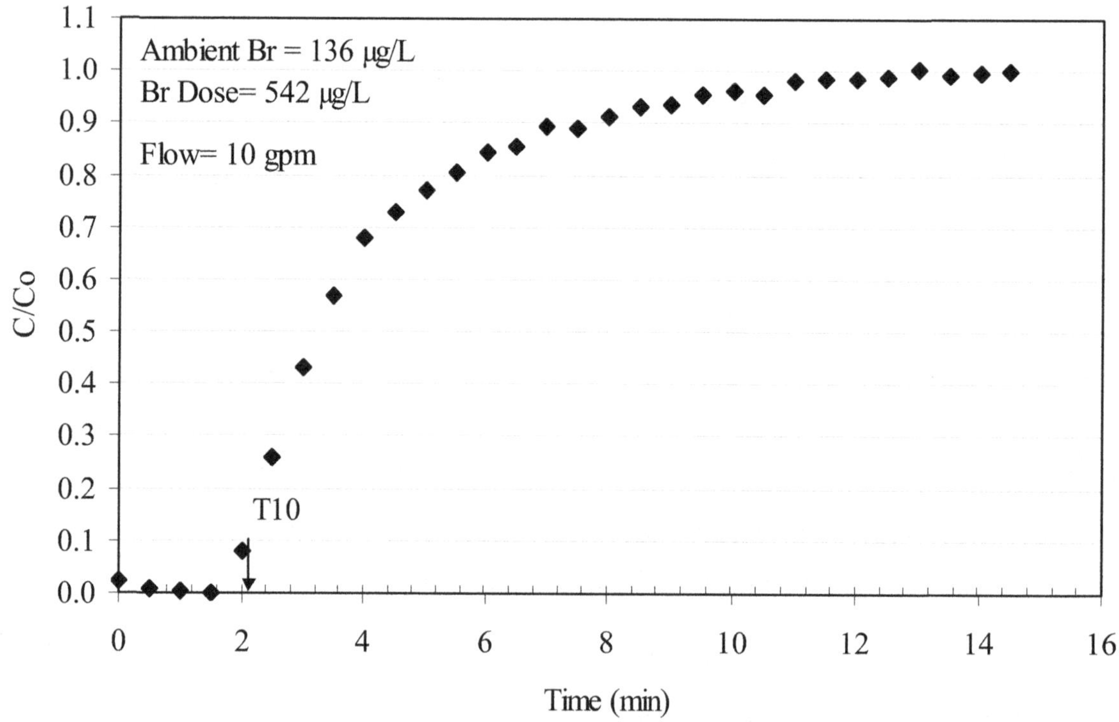

Figure 4.3 Tracer test curve conducted at 10 gpm flowrate

Low Power/High HRT Testing

Several electrolytic reactor runs were conducted to assess the acceptable range of operations for bromide removal and control of DBPFP. Variables included the HRT, raw water bromide concentration, and applied levels of current and power. The runs were conducted under low power and low flowrate conditions. Table 4.1 summarizes the experimental conditions applied during this phase of testing.

Effect of Initial Bromide Concentration

Figure 4.4 illustrates the result for the first set of experiments, conducted under an HRT of 11.2 min, ambient bromide concentration of 140 μg/L, and under current values of 0 to 35 amps (0 to 949 watts). In Figure 4.4, total bromide is the sum of bromide anion in the water and the bromine in the water. A decrease in overall total bromide is observed only when two reactions take place: bromide oxidizes to bromine and bromine volatilizes. It appears that there is no limit on how much bromide can be oxidized to bromine. Most of the bromide anion was oxidized to bromine in the aqueous phase.

Bromide oxidation appeared to be directly proportional to the applied current. This follows the principles of Faraday's Law (Chapter 3), i.e., an increase in the applied current results in a higher rate of reaction. The decrease in bromide ion concentration per 1 amp of applied current ranged from 2.3 to 4.2 μg/L per amp of applied current. The optimal removal occurred at 15 amps or approximately 200 watts (refer to bolded values in Table 4.2). At this point, the highest bromide removal was measured with the least amount of free available bromine accumulation. It appears that bromine cannot be volatilized below a certain threshold concentration of bromide ion in the water (i.e., about 50 μg/L). Below that level, an accumulation in bromine levels in the aqueous phase was observed, resulting in a plateau. The level of bromine accumulation needs to be minimized, since upon chlorination, bromine is more reactive than chlorine to form DBPs (Singer 1999).

Table 4.1
Experimental conditions applied during low power/high HRT testing

Parameter	Value	Additional comments
Flowrate range	2 to 7 gpm	
Theoretical HRT	3.2 to 11.2 min	
Current amperage	0 to 35 amps	
Power	0 to 950 watts	
Raw water bromide	134 to 150 μg/L	Ambient levels
Raw water bromide	435 to 912 μg/L	Spiked levels after addition of bromide

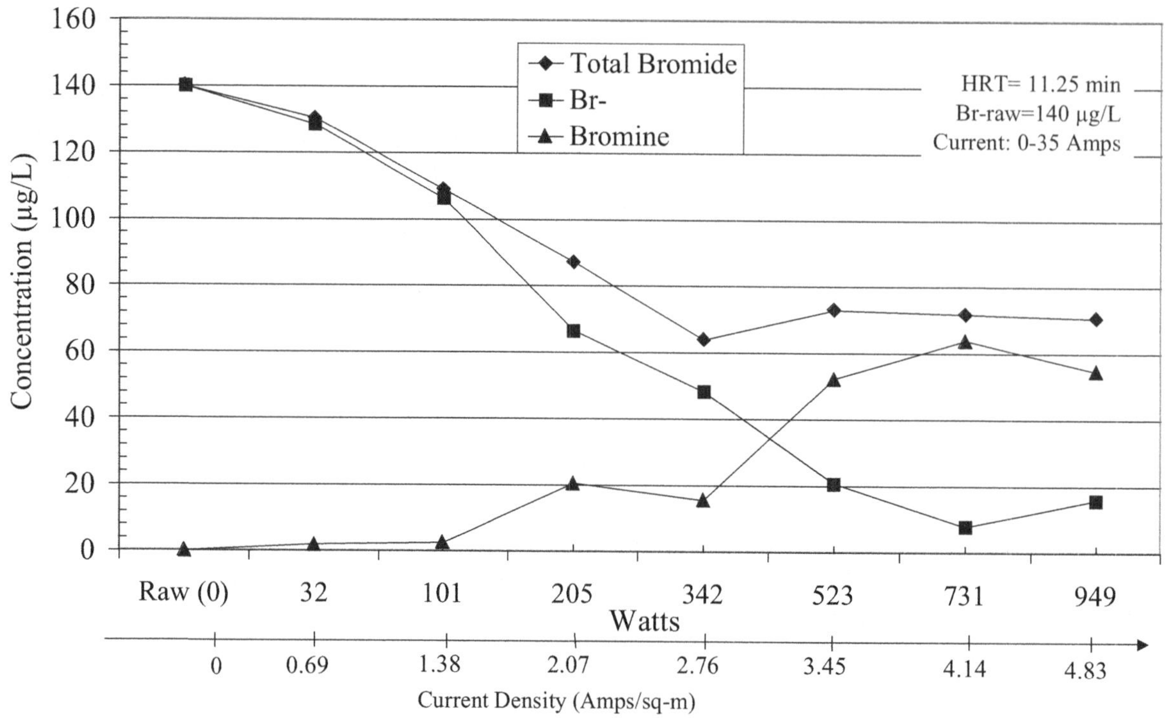

Figure 4.4 Bromide oxidation under ambient bromide and high HRT conditions

Table 4.2
Effect of initial bromide concentration on bromide removal and bromine accumulation under high HRT condition(11.2 min)

Current (amps)	5	10	15	25	35
Current density (amps/sq-m)	0.69	1.38	2.07	3.45	4.83
Power range (watts)	30 to 32	97 to 101	200 to 205	510 to 523	949 to 955
Total bromide removal per amp (µg/L/A)					
Ambient bromide (i.e. 150 µg/L)	3.9	4.1	**4.2**	3.1	2.3
Spiked bromide (435 µg/L)	43.5	48.2	**55.5**	60	52.5
Bromine accumulation per amp (µg/L/A)					
Ambient bromide (i.e. 150 µg/L)	0.4	0.3	**1.4**	2.1	1.6
Spiked bromide (435 µg/L)	0.4	0.9	**1.6**	3.1	3.8

Note: Bold values indicate conditions associated with highest percent of bromide removal and lowest accumulation of aqueous bromine

Since the raw water quality fluctuates seasonally, the experiment was repeated by spiking bromide to a target concentration of 435 µg/L. The same trends in the bromide and bromine curves were observed as in the ambient bromide experiment (Figure 4.5). A threshold in bromide removal was observed when the bromide concentration decreased to 60 µg/L as a result of electrolysis. Below that concentration, bromine accumulation appeared to occur. The optimal zone of testing was estimated at 15 amps or 200 watts, equivalent to the optimal zone of electrolysis during the ambient bromide experiment.

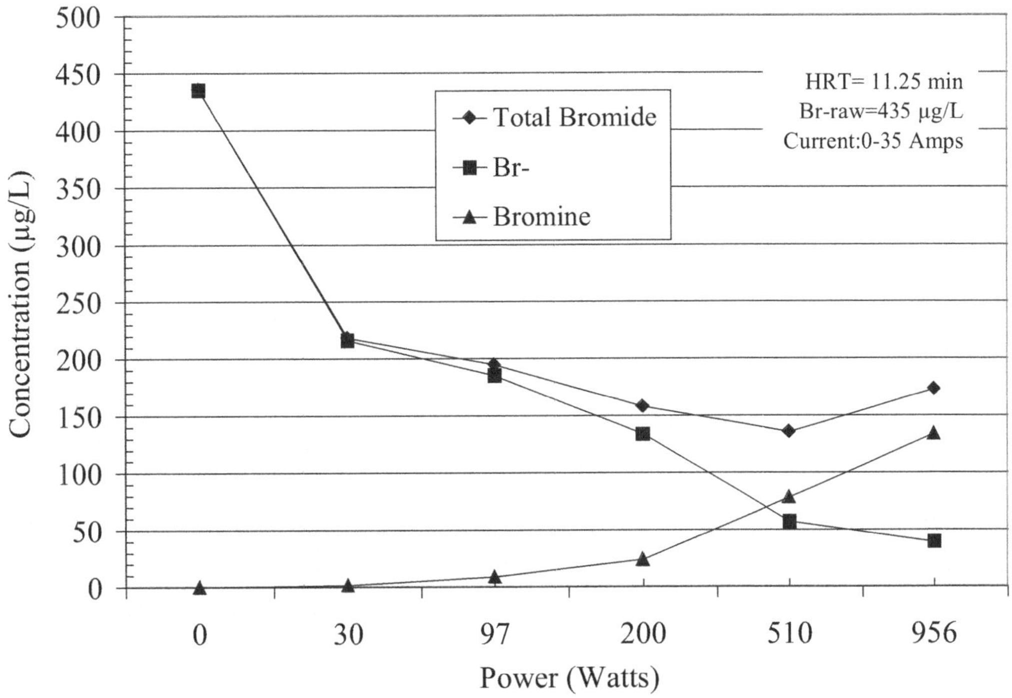

Figure 4.5 Bromide oxidation under spiked bromide and high HRT conditions

Percent bromide removals and bromine fractions under ambient and spiked bromide conditions are illustrated in Figure 4.6 and presented in Table 4.2. While the percent bromide removal and the bromide removal per amp were higher in the spiked bromide samples, the bromine to bromide fraction and bromine accumulation were equivalent in both sets of samples. Therefore, the initial bromide concentration did not appear to have an effect on the bromide oxidation process.

Effect of Hydraulic Residence Time

The effect of hydraulic residence time and of mixing on the bromide electrolysis process was investigated by varying the flowrate through the reactor. During this initial phase of testing, the highest tested flowrate was limited to 7 gpm, equivalent to 3.2 min HRT, and the highest applied current was 35-amps. The total bromide, aqueous bromide anion, and bromine curves are included in Appendix A (total bromide is the sum of aqueous bromide and bromine). Figure 4.7 illustrates the percent bromide removal and percent of bromine accumulated as a function of current and HRT/mixing conditions. The bromide oxidation to bromine was not as effective at 3.2 min HRT as compared to 11.2 min HRT. The reaction is rate limited, and for the reaction to occur at lower residence times it would require a higher applied power. In addition, as discussed in Chapter 3, mixing was demonstrated to negatively impact the bromide oxidation reaction. Thus, under the higher flowrates tested, sub-optimal conditions for bromide oxidation exist because of low HRTs and the presence of mixing.

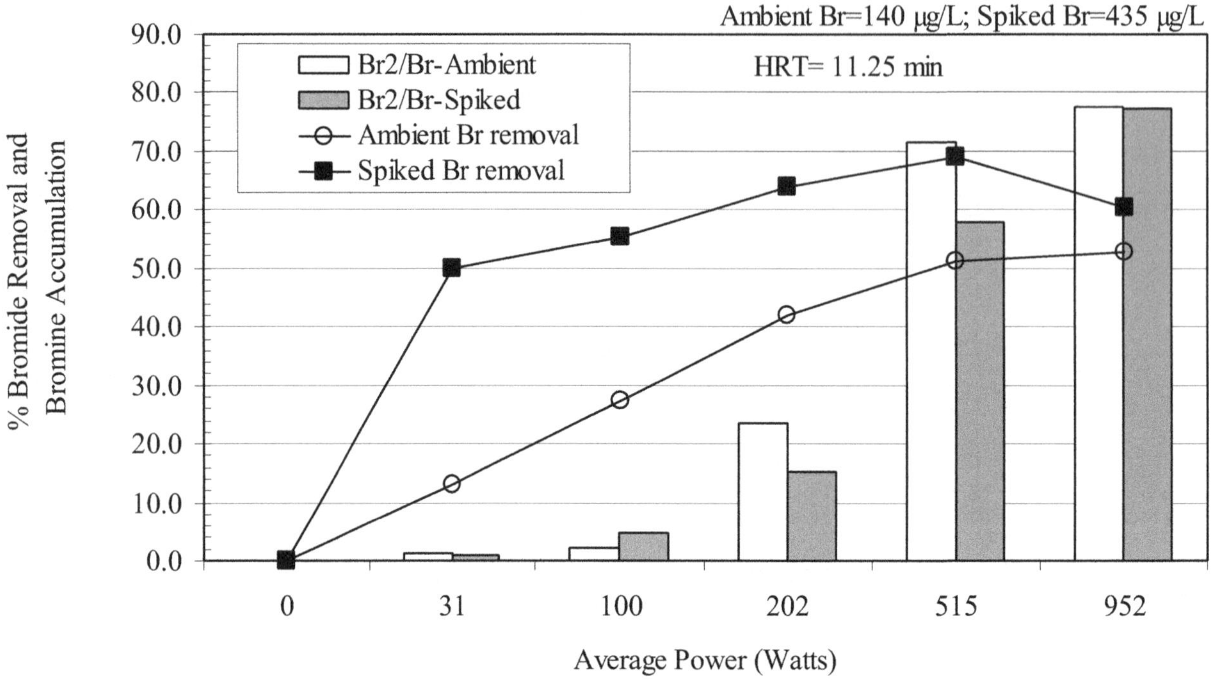

Figure 4.6 Percent bromide removal and bromine accumulation under ambient and high bromide conditions

Bromide removal per amp and bromine accumulation per amp are summarized in Table 4.3 for each HRT condition. The bolded values under the tested conditions in Table 4.3 were assessed as optimal for maximizing bromide removal and minimizing residual bromine. The optimal current ranges under the 5.6-min and 11.2-min HRTs lie somewhere between 15 to 25 amps, and 10 to 15 amps, respectively. It should be noted that under the 3.2-min HRT condition, optimal conditions for bromide removal were not achieved (i.e., the bromide oxidation was not effective). A higher power would likely be required.

High Power/Low HRT Testing

To compliment the above findings, a series of experiments were conducted to investigate the effect of high power and high flow/low HRT on the bromide oxidation process. Three levels of current and four concentrations of bromide were tested under HRTs of 1.5, 2.2, and 4.5 minutes. A typical plot of total bromide, aqueous bromide, and bromine is illustrated in Figure 4.8. Although the higher applied power activated the oxidation reaction of bromide into bromine under low HRT conditions, and although at higher power there is a high production of oxygen bubble (Figure 4.9), an accumulation of residual bromine was observed. The occurrence of turbulence within the reactor zone is likely detrimental for the oxidation reaction. Total bromide, aqueous bromide, and bromine curves for all other conditions are included in Appendix B.

Table 4.3

Effect of HRT condition on μg/L bromide removal and bromine accumulation

Current (Amps)	5	10	15	25	35
Current density (Amps/sq-m)	0.69	1.38	2.07	3.45	4.83
Power range (Watts)	30-32	97-101	200-205	510-523	949-955
Total bromide removal per amp (μg/L Br removed/A)					
3.2 min; Br=912 μg/L	7.69	5.93	5.04	4.96	5.17
5.6 min; Br=134 μg/L	0.1	0.6	0.6	**1.2**	1.3
5.6 min; Br=630 μg/L	1.7	outlier	2.5	**3.8**	3.6
11.2 min; Br= 150 μg/L	3.9	4.1	**4.2**	3.1	2.3
11.2 min; Br= 435 μg/L	**43.4**	24	18.5	**12**	7.5
Bromine accumulation per amp (μg/L/A)					
3.2 min; Br=912 μg/L	0	0.2	0.5	1.1	2.4
5.6 min; Br=134 μg/L	0.7	0.04	**0.6**	1.8	1.4
5.6 min; Br=630 μg/L	0	0	**0.6**	2.6	3.9
11.2 min; Br= 150 μg/L	0.4	0.3	**1.4**	2.1	1.6
11.2 min; Br= 435 μg/L	0.4	0.9	**1.6**	3.1	3.8

Note: Bold values indicate conditions associated with highest percent of bromide removal and lowest accumulation of aqueous bromine

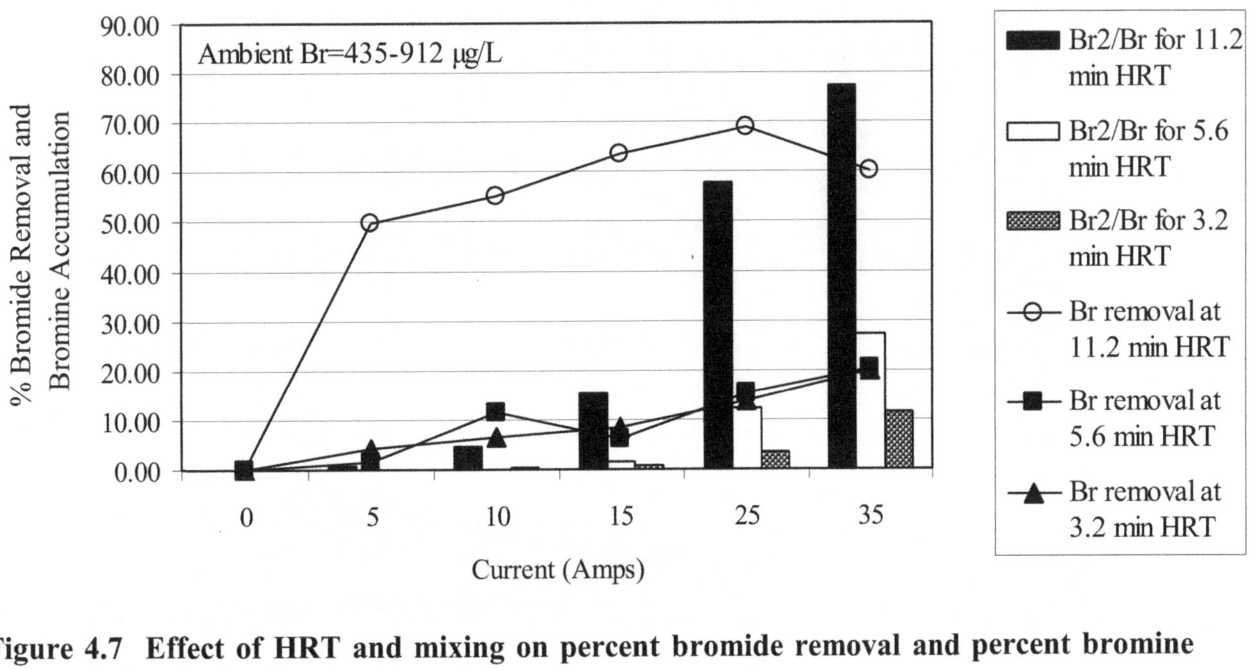

Figure 4.7 Effect of HRT and mixing on percent bromide removal and percent bromine accumulation during low power testing

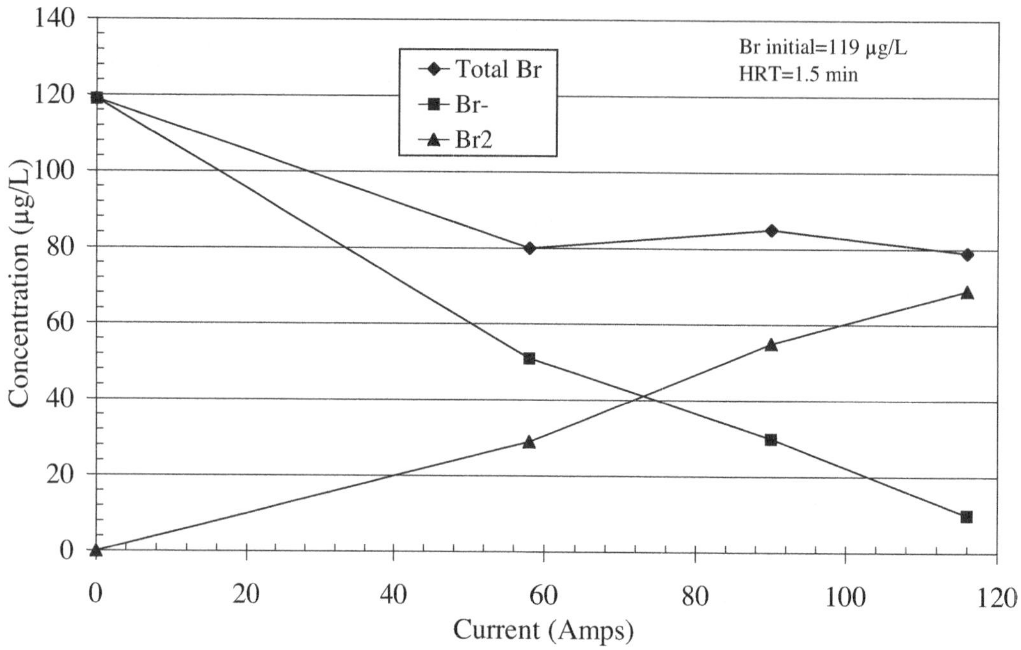

Figure 4.8 Typical bromide and bromine curves under high power and low HRT condition

Figure 4.9 Oxygen bubbles formed during high-power electrolysis testing (i.e., 120 amps at 2 gpm flowrate)

Bromide removals and bromine accumulations for all conditions are included in Appendix C. Optimal ranges of operation were identified for each HRT/flowrate condition. Based on the combined bromide removal and bromine accumulation results, the optimal range of current was identified as 60 amps or less for HRTs of 2.2 to 4.5 min, and 60 to 90 amps for the 1.5-min HRT. The limiting factor is the accumulation of bromine and the occurrence of turbulence under the higher flowrate. The combined results from the low and high HRT testing are presented in Table 4.4. As can be seen, there are data gaps that would need to be filled to refine the optimal operational conditions. Figure 4.10 summarizes optimal power conditions under all tested HRT conditions, required to achieve maximum bromide removal with minimum bromine accumulation.

Table 4.4
Optimal current and power conditions for bromide removal

HRT	Initial bromide concentration (μg/L)	Current (amps)	Power (kW)
1.5 min	119-677	58-90	2.4-5.8
2.2 min	115-578	≤60	≤2.6
3.2 min	912	>35 but <60	>1.0 but <2.6
4.5 min	121-335	> 25 but <60	> 0.5 but <2.6
5.6 min	134-630	15-25	0.2-0.5
11.2 min	150-435	10-15	0.1-0.2

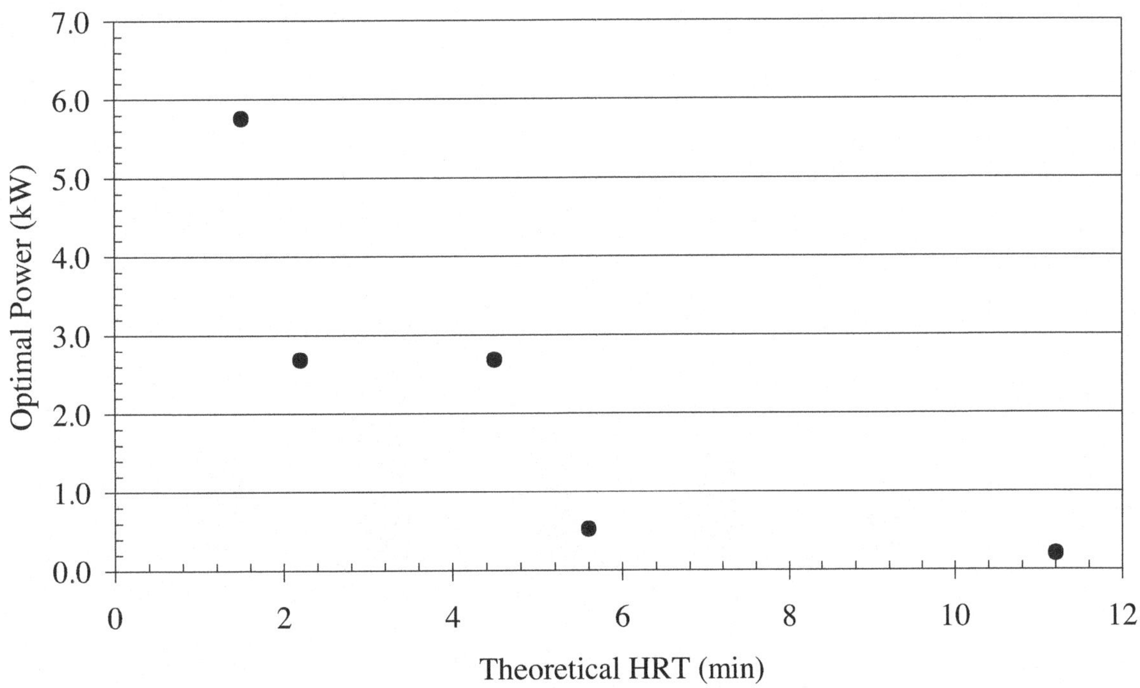

Figure 4.10 Optimal power versus HRT for bromide removal

It should be noted that optimal conditions attained under a low HRT are not equivalent to optimal conditions under a high HRT. The potential benefits in terms of bromide removal (and subsequently DBPFP reduction) associated with each HRT/flow condition will have to be assessed in terms of capital and power costs. To illustrate this point, the percent bromide removals and bromide removals measured under optimal conditions of power were plotted for each HRT condition (Figure 4.11). Even under optimal conditions for bromide oxidation, an 11.2-min HRT resulted in higher bromide removed per amp of applied current as compared to HRTs of 5.6 minutes and lower. The process is limited by the accumulated fraction of bromine, which lowers the overall total bromide removal. The process is also limited by the presence of turbulence within the reaction zone under the higher flowrates.

While overall bromide removal is important, lowering the DBPFP is the ultimate goal. The direct effect of electrolysis on DBPFP is investigated in subsequent sections.

Effect of Reactor Depth

An experiment was conducted to assess the bromide oxidation over the depth of the DSA plates. The plates depth is 3 ft (91.4 cm), and samples were collected for bromine analysis at 0 cm (bottom of plate), 5 cm, 10 cm, 50 cm, and 91 cm (top of plates). The sampling device (siphon system) was inserted in between two of the fourteen plates. The experiment was conducted at ambient bromide concentration (152 μg/L), 2-gpm flowrate, and 35 amps of current.

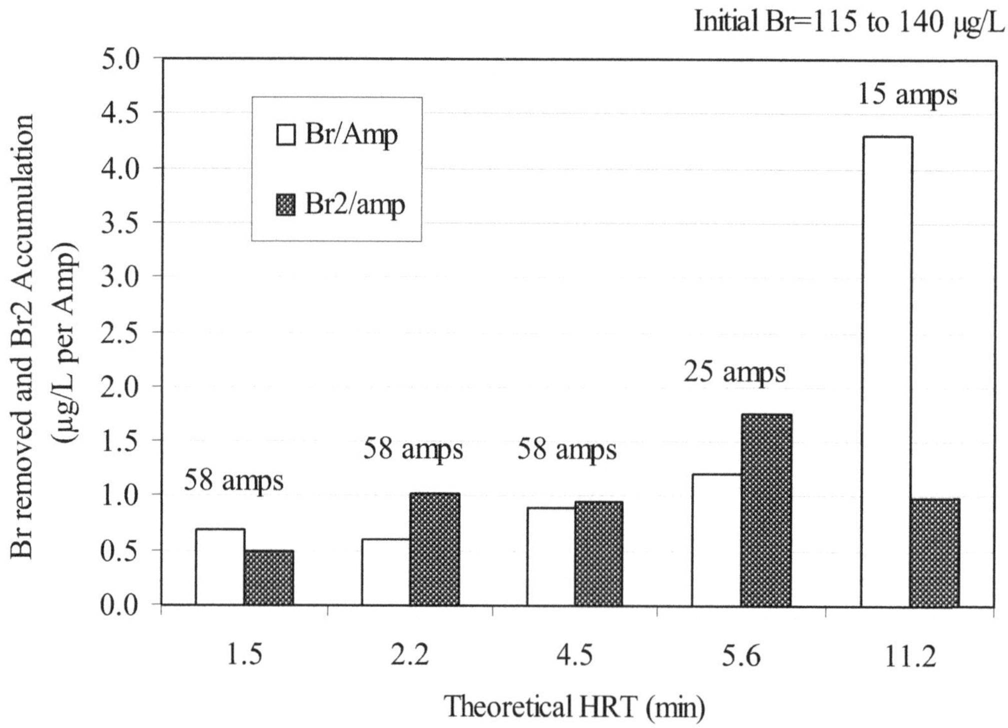

Figure 4.11 Impact of HRT on overall bromide removal

A critical finding is that the majority of the bromide removal, i.e., 152 μg/L down to the threshold level of 75 μg/L, occurred in the first 10 cm of depth. Beyond that depth, bromine accumulation appears to be taking place. Similar results were obtained in a previous experiment conducted by Kimbrough and Suffet (unpublished results). In that experiment, bromide was completely oxidized to bromine in the first 3 cm of a 32-cm deep cylindrical reactor.

Although limited data are available pertaining to this subject, the ramifications can be extremely important. Under the tested conditions of 2-gpm flowrate, the required HRT would drop from 11 minutes to 1.2 minutes, which would significantly decrease the required footprint. volume requirements, and required number of DSA plates. In addition, bromine accumulation would be minimized, which could decrease the DBPFPs. This is illustrated in Figure 4.12.

Despite the fact that above results obtained at 15 gpm flowrate (1.5-min HRT) resulted in sub-optimal bromide removal, the occurrence of turbulence and high velocity within the reaction zone under that flow condition may have negatively impacted the bromide oxidation reaction. This concurs with the findings of mechanistic effects testing presented in Chapter 3. Thus an HRT of 1 to 1.5 min measured under low turbulence/velocity conditions created a better environment for bromide oxidation as opposed to the same HRT under high flow conditions.

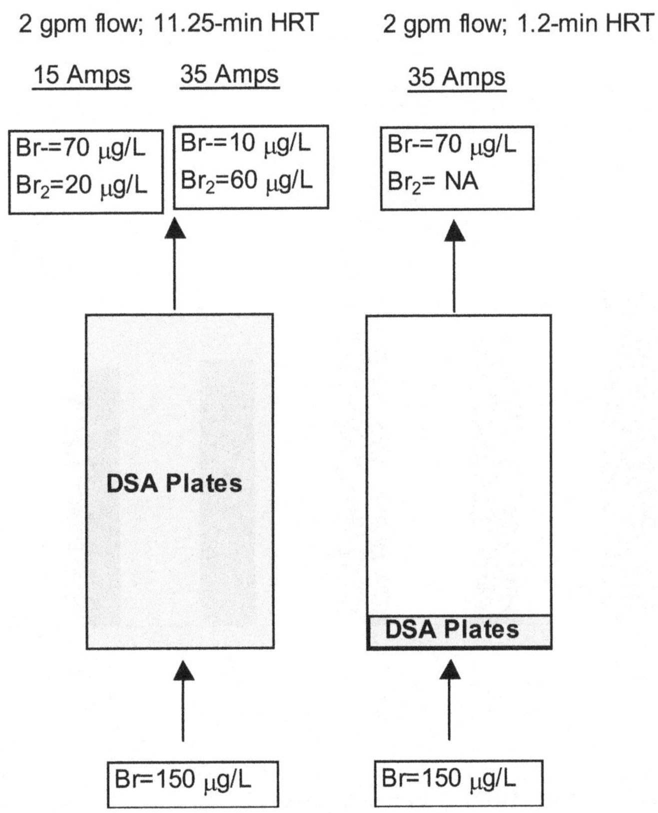

Figure 4.12 Schematic of effect of DSA plate depth on bromide removal

REACTOR TESTING FOR DBPFP REDUCTION

The electrolysis process was also investigated in terms of its effect on DBPFP. Experimental variables included HRT, initial bromide concentration, and the effect of dechlorination.

Chlorine Generation

As discussed in Chapter 3, in addition to bromide being oxidized into bromine during electrolysis, chlorine is formed at the anode as a direct result of oxidation of the chloride anion present in the raw water ($E_o = +1.36$ V). Figure 4.13 is a typical curve of chlorine formation as a direct result of applying current. As expected, the chloride concentration was observed to decrease as a direct function of the applied current, and as a result, chlorine was formed. However, the level of chlorine formation in the electrolyzed samples was not as high as anticipated. The highest chlorine concentration was measured at 1.5 mg/L at 30 amps. Chloride oxidation into chlorine is likely to increase at higher applied amps. As a note, although only limited data are available on chlorine levels under higher amperage conditions, a 6-mg/L free available chlorine level was measured when 120 amps (9.8 kW power) were applied.

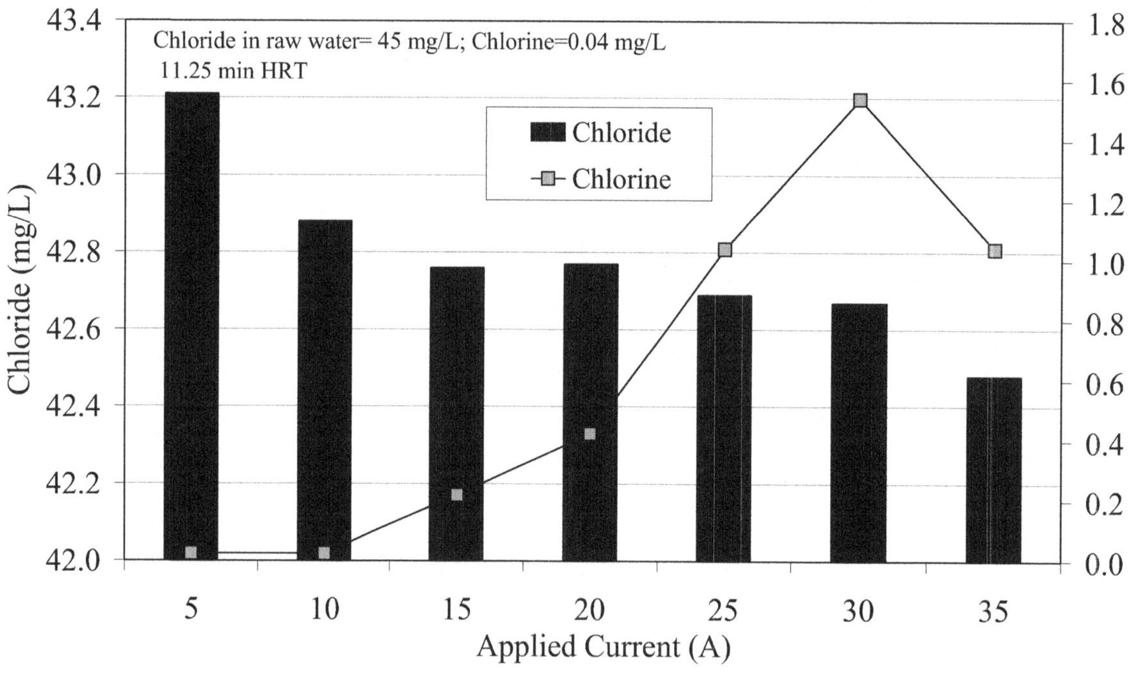

Figure 4.13 Chlorine generation during electrolysis

Instantaneous DBP Formation

As a result of bromine accumulation and chlorine formation, THMs, HAAs, and TOXs were formed in electrolyzed samples, increasing with increasing current (Figure 4.14). These were measured on-site in all electrolyzed samples, with no quenching of the chlorine/bromine formed as a result of electrolysis. TOX increased from 12 to 141 μg/L as a result of applying 35 amps through the water. Similarly, THMs and HAAs increased to measurable levels. While this may appear counter-productive, previous research has shown that the removal of bromide anion decreased the formation potential of brominated THMs and HAAs. This is investigated in the next section.

Interestingly, equivalent levels of THMs were formed in ambient and bromide-spiked samples (Figure 4.15). This is likely due to equivalent chlorine and bromine concentrations formed in both sets of samples. It is noted that the experiment did not test natural wasters, which contain varying concentrations of salts and minerals (e.g., chloride, sulfate, bromide). To truly assess the effect of bromide on DBP formation following electrolysis, the experiment should be repeated using natural waters with varying levels of bromide.

DBP Formation Potentials

During the first part of the project, DBPFP tests were conducted on raw and electrolyzed samples using high free chlorine target doses (10 mg/L) and with no quenching of the instantaneous chlorine/bromine formed as a result of electrolysis.

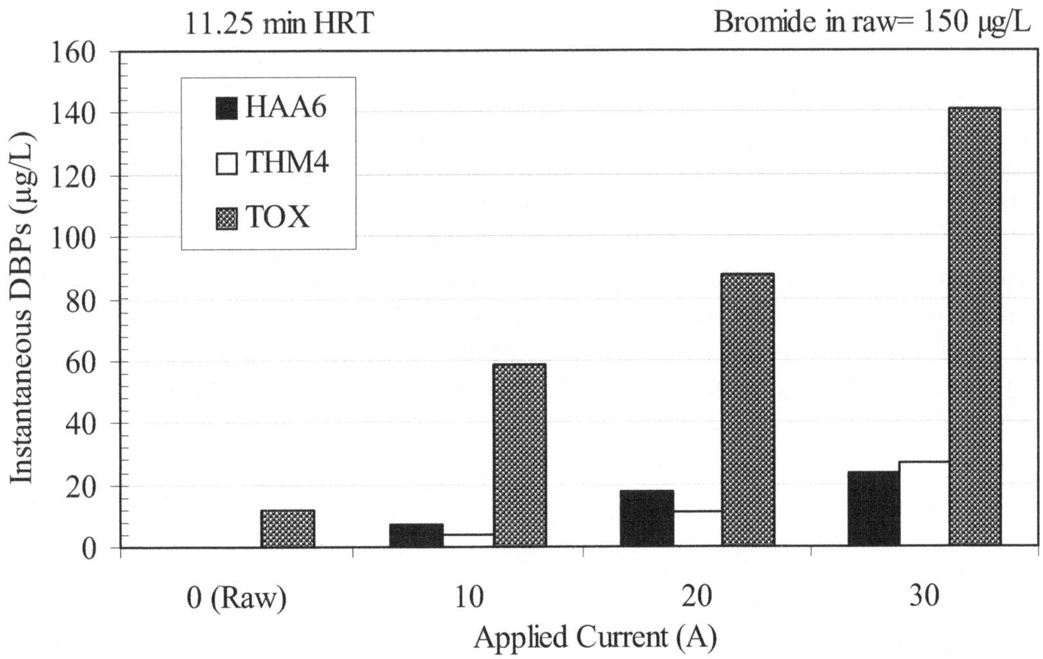

Figure 4.14 Instantaneous DBP levels as a direct result of electrolysis

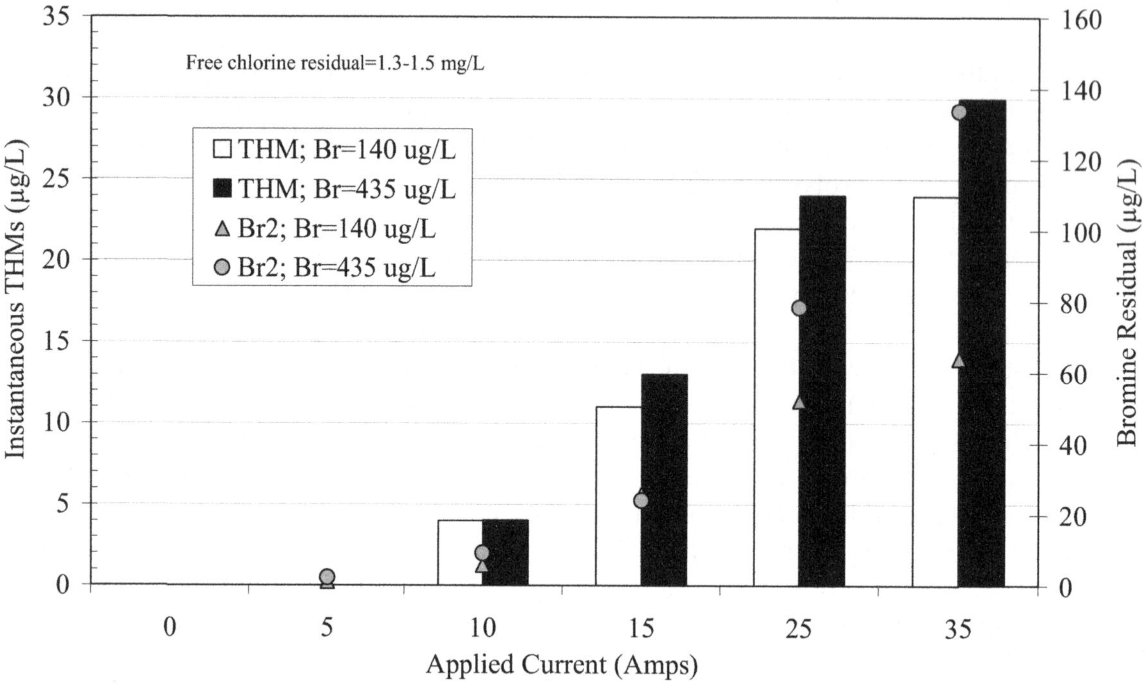

Figure 4.15 Effect of initial bromide concentration on THM levels formed as a direct result of electrolysis

Ambient Bromide Conditions

In this experiment, the DBPFP were measured under ambient bromide conditions (i.e., initial bromide concentration of 140 µg/L) and at a high HRT of 11.2 minutes. Figure 4.16 illustrates the bromine residuals, the instantaneous THMs formed, and the formation potentials of THM species in raw and electrolyzed samples after 24-hour of incubation. A 10 percent decrease in THMFP was observed in electrolyzed samples as compared to the raw water sample, when 10 to 20 amps of current (i.e., 100 to 340 watts of power) were applied. Higher currents were observed to be detrimental; i.e., had higher THMFP. This is likely due to the high residual bromine in those samples. On the other hand, a current amperage of 20 A (i.e., 340 watts) was observed to decrease HAA6 formation potentials (HAA6FPs) by 12 percent (Figure 4.17).

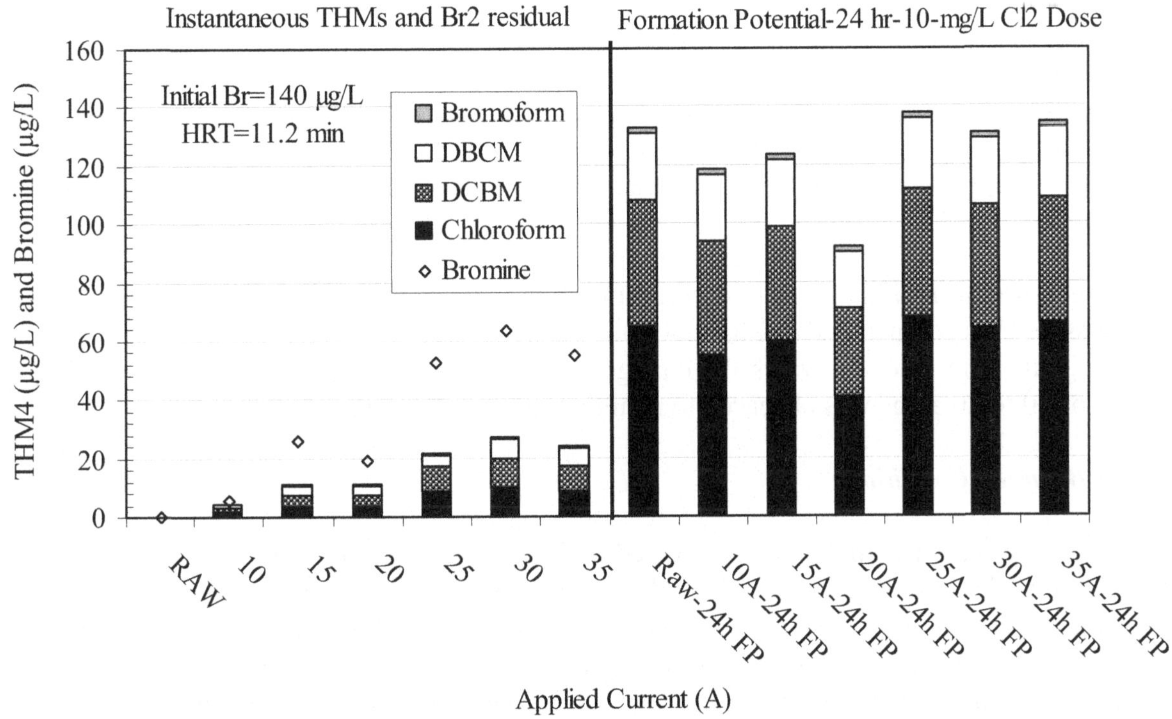

Figure 4.16 Instantaneous THMs and THM4 formation potentials under ambient bromide conditions

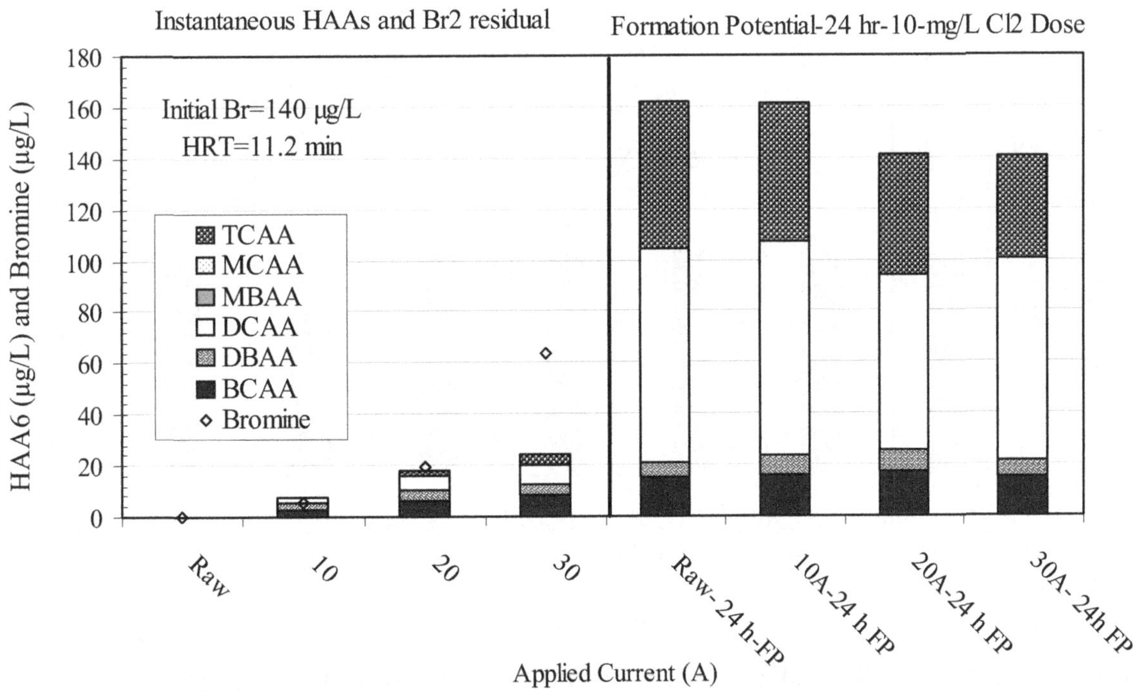

Figure 4.17 Instantaneous HAAs and HAA6 formation potentials under ambient bromide conditions

To further investigate this point, the bromine incorporation factor (n) under ambient bromide conditions was computed and illustrated in Figure 4.18. This factor (n) is the molar concentration of bromine in the THMs or HAAs divided by the molar THM4 or HAA6 concentration. As the current increases and bromide is removed with minimal bromine accumulation, a drop in n is expected. In Figure 4.18, the factor n was observed to decrease in HAAs under a current of up to 20 amps. It was observed to increase when a higher current of 30 amps was applied. It should be noted that only five of the nine HAA species were measured in this experiment, and the effect on overall brominated HAA species would have to be confirmed. In terms of bromine incorporation in the THMs, no trend was observed as a function of applied current. Estimating n under high bromide spike conditions may provide more insight in this issue. This is illustrated in Figure 4.20.

Figure 4.18 also illustrates the change in TOX concentration with applied amperage. A current of 10 to 20 amps was observed to result in an 8 percent decrease.

Spiked Bromide Conditions

The above experiments were repeated under spiked bromide conditions to verify whether a higher initial bromide concentration plays a role in the DBP formation potentials. In this experiment, bromide was spiked to a target concentration of 435 μg/L and the HRT was set at 11.2 minutes. In Figure 4.19, the 24-hr THMFPs were observed to decrease by 30 percent as a result of applying a direct current. The optimal current conditions were 15 amps, equivalent to 200 watts of power. Beyond that, THMFPs increased. This again is likely due to the bromine accumulation in the electrolyzed samples, especially at currents exceeding 15 amps.

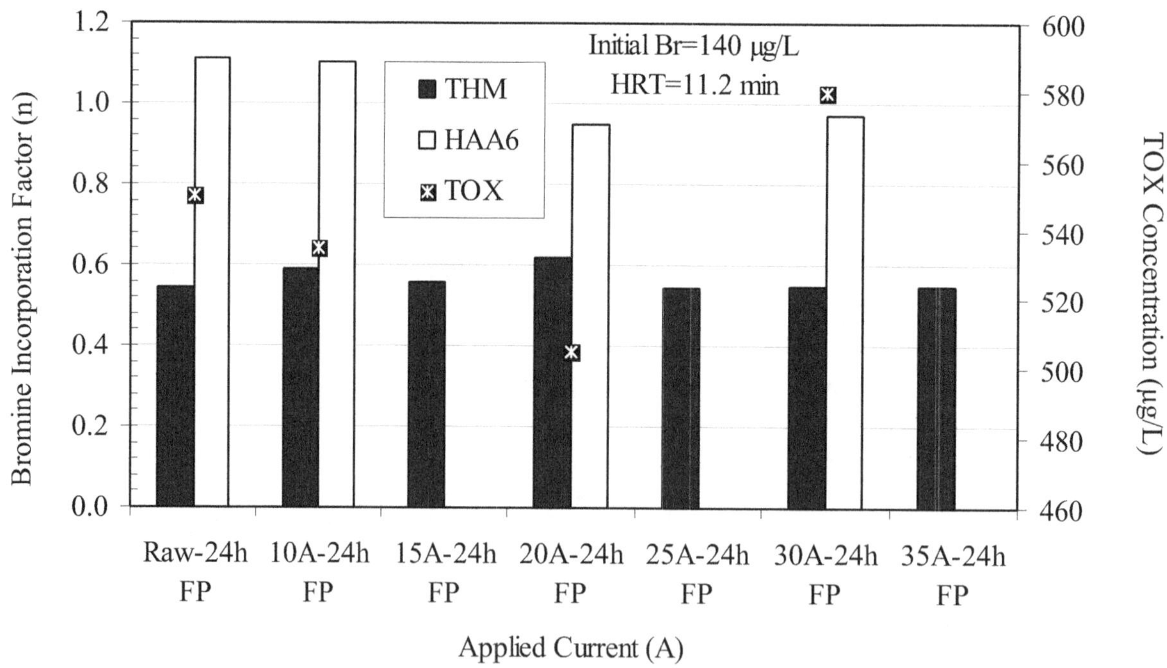

Figure 4.18 Effect of electrolysis on the bromine incorporation factor (n) in THMs and HAAs and on the TOX concentration under ambient bromide conditions

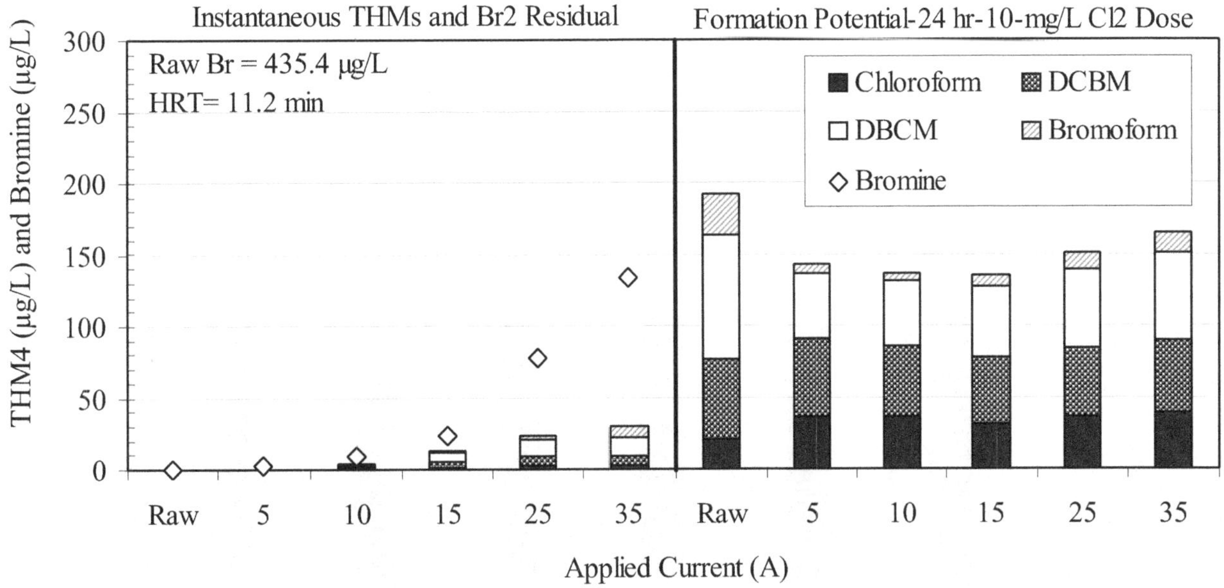

Figure 4.19 Instantaneous THMs and THM4 formation potentials under spiked bromide conditions

The bromine incorporation factor (n) under spiked bromide conditions was computed and illustrated in Figure 4.20. As the current increases and bromide is removed with minimal bromine accumulation, a drop in n is expected. This was observed in Figure 4.20, where the n in THMs decreased from 1.44 in the raw water sample to 0.95 in the sample electrolyzed under a 10-amp current, and decreased from 0.87 to 0.5 in the HAAs.

Figure 4.20 also illustrates the change in TOX concentration with applied amperage. A 5 percent decrease in TOX formation potential (TOXFP) was observed upon application of 5 to 10 amps of current. An increase was observed beyond those amperages.

Effect of HRT and Mixing

In the previous discussion on bromide removal, an 11.2-min HRT calculated under a 2 gpm flowrate was assessed as optimal in terms of bromide removed per amp applied. To assess the optimal HRT for DBP reduction, a series of formation potential tests were conducted under variable flowrate conditions. It should be noted that these tests were limited to low flow/high HRT conditions, and tests would have to be repeated under low HRT (i.e., flowrates of 5 mg/L and higher) conditions. The DBPFP bars plots are included in Appendix D. The decrease in THMFP was assessed under three HRT conditions (Figure 4.21). The shown decrease in THMFP is associated with the optimal current condition. The removal of THMFP was approximately five times greater at 11.2-min HRT and low mixing condition as compared to lower HRTs. Unfortunately, limited data are available to confirm these results. No data are available for HAAs and TOX.

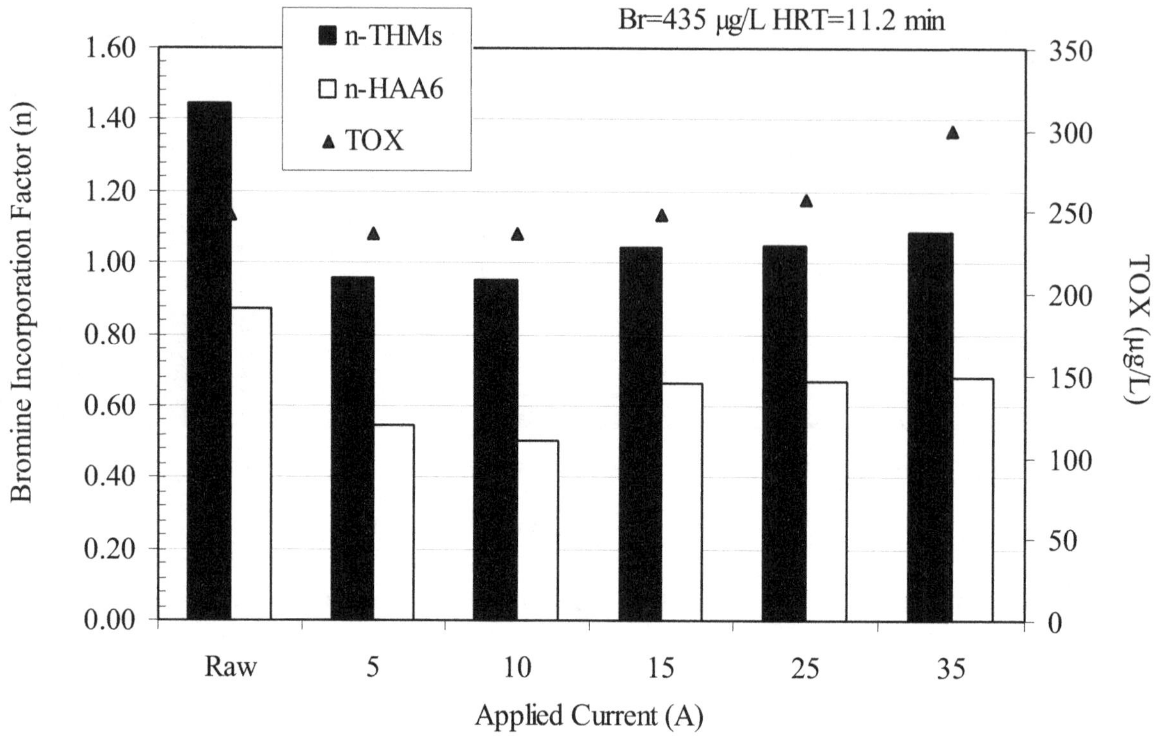

Figure 4.20 Effect of electrolysis on the bromine incorporation factor (n) in THMs and HAAs and on the TOX concentration under high bromide conditions

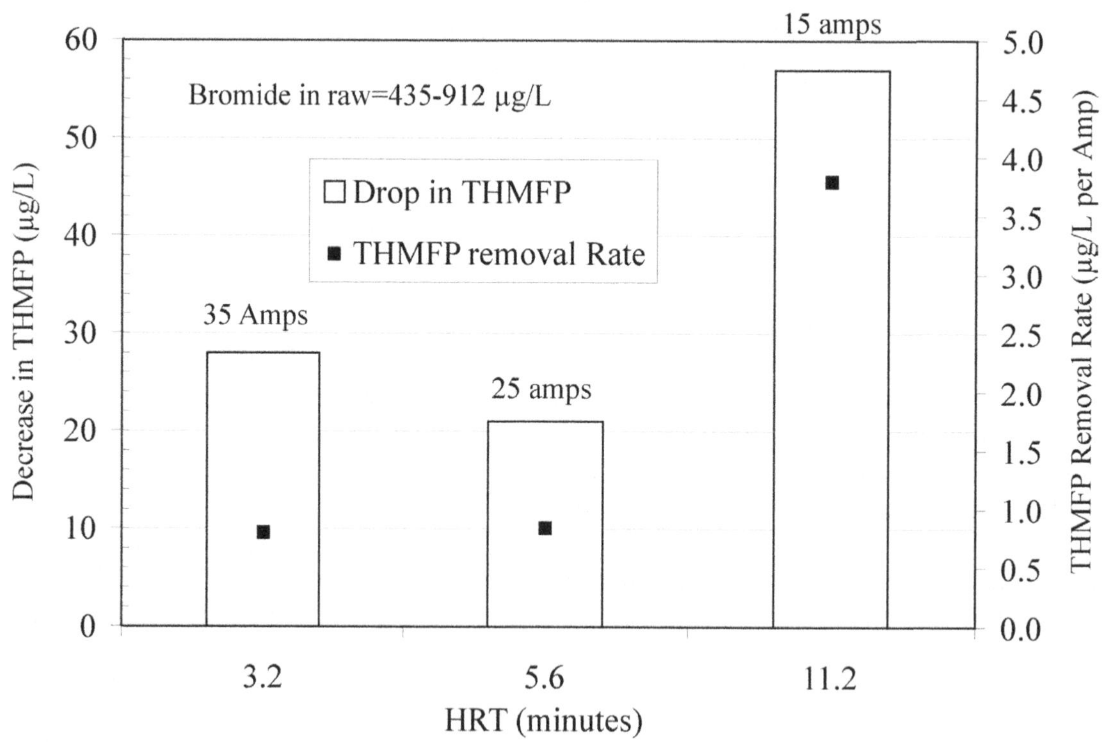

Figure 4.21 Impact of HRT on THMFP

Effect of Dehalogenation

The occurrence of chlorine and bromine in the electrolyzed samples appears to be responsible for some of the high DBPFPs, especially at the higher power levels. The volatilization of bromine appeared to be hindered and limited by a minimum threshold bromide concentration in the water. To further qualify the role of bromine and chlorine in the formation potentials of DBPs, a series of experiments were conducted by adding a dehalogenating agent to the electrolyzed samples, prior to the addition of free chlorine for DBPFP tests. These tests may thus give indication of the electrolytic reactor performance in low chloride/bromide waters, where a low chlorine/bromine residual would presumably form. Sodium thiosulfate was used for that purpose. By adding a reducing agent, the reaction of accumulated aqueous bromine and chlorine bromine for formation of DBPs is interrupted. The experimental conditions are presented in Table 4.5.

Dehalogenating the electrolyzed samples appeared to decrease the THMFPs (Figure 4.22). A small decrease in formation potential (11 µg/L) was observed as a result of applying 15 A of current. This decrease doubled to 23 µg/L when a dehalogenting agent was added to the samples.

To assess the effect of dechlorination under lower chlorine formation potential conditions, a subsequent THMFP test was conducted in which the following conditions were used:

- Three samples were electrolyzed under applied power intensities of 225 watts, 550 watts, and 1015 watts (i.e., 15 amps, 25 amps, and 35 amps).
- The chlorine residuals in those samples were respectively 0.6, 1.1, and 1.6 mg/L. The instantaneous THMs were measured in the three samples.
- Free chlorine was added to raw/unelectrolyzed water samples to match the above residuals.
- Each of the three electrolyzed samples was split in two halves. Only one of the halves received a dehalogenating agent and was then re-chlorinated to achieve the target free chlorine residual. The other half was directly rechlorinated.
- The chlorinated raw, the electrolyzed, and the dechlorinated/rechlorinated electrolyzed samples were incubated over a 24-hr period.

Table 4.5
Formation potential conditions during dechlorination experiments

Initial Br^- concentration	117 µg/L
Theoretical hydraulic retention time	11.2 min
Current range	0 to 35Amps
Power range	0 to 950 Watts
Free chlorine dose	10 mg/L
Incubation period	24 h

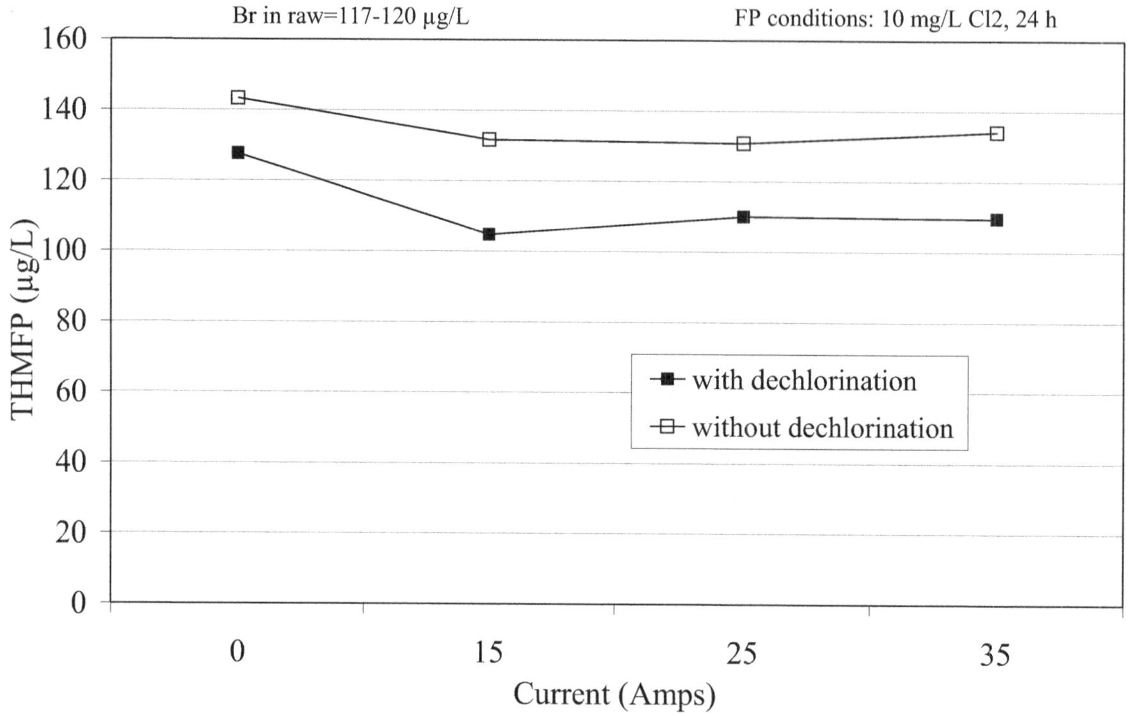

Figure 4.22 Effect of dehalogenation on THMFP in electrolyzed samples

Results are illustrated in Figure 4.23. Electrolyzing the water samples appeared to increase the THMFPs. The power setting may have been non-optimal in terms of bromide removal and bromine accumulation. However, under two of the three power conditions, dechlorinating and rechlorinating the electrolyzed samples decreased the THMFP to levels lower than those in the raw water samples. When comparing THM species in electrolyzed and dechlorinated/rechlorinated electrolyzed samples, a drop is observed in the chloroform, dichlorobromomethane, and dibromochloromethane species.

These data are somewhat limited, and more research is required to confirm the findings. The addition of hydrogen peroxide as a quenching agent should be tested since dechlorination would likely occur at a faster rate as compared to sodium thiosulfate.

TOBr and TOCl Formation Potentials

In addition to THMFP and HAA formation potential (HAAFP), speciated TOX (i.e., TOBr and TOCl) formation potentials were measured in raw and electrolyzed samples and under ambient and bromide-spiked conditions. The application of current appeared to decrease the TOBr and increase the TOCl. This was observed in both sets of samples (Figures 4.24 and4.25). Further research is required to confirm these findings.

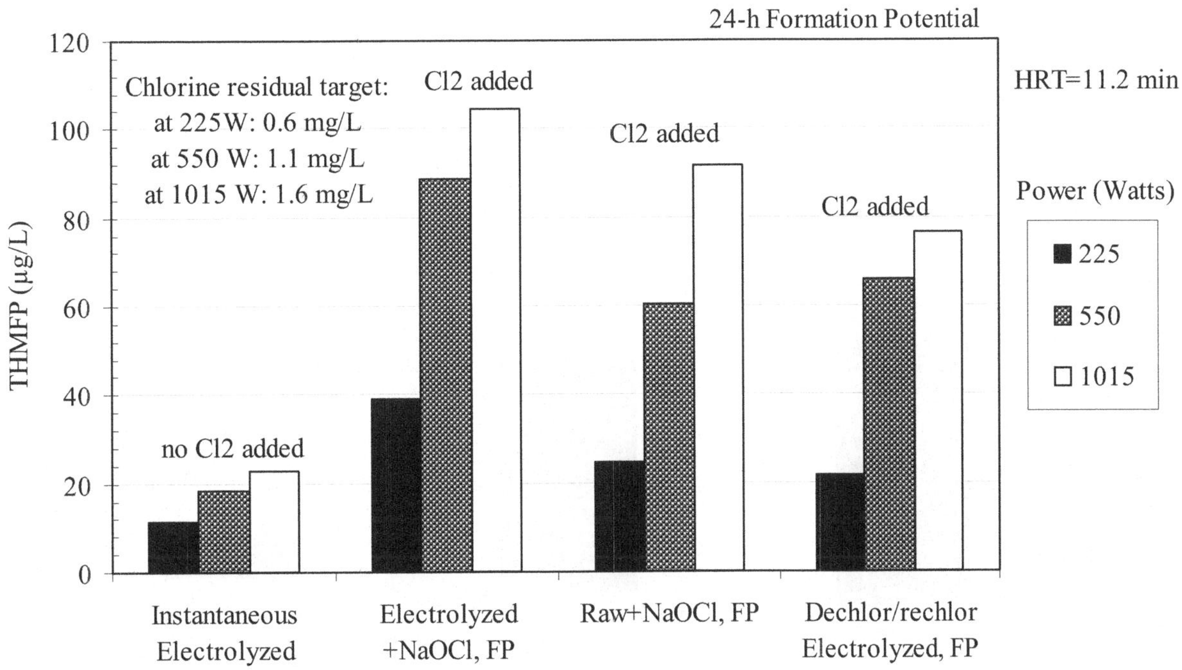

Figure 4.23 Effect of dechlorination/debromination on THMFP in electrolyzed samples and low chlorine formation potential conditions

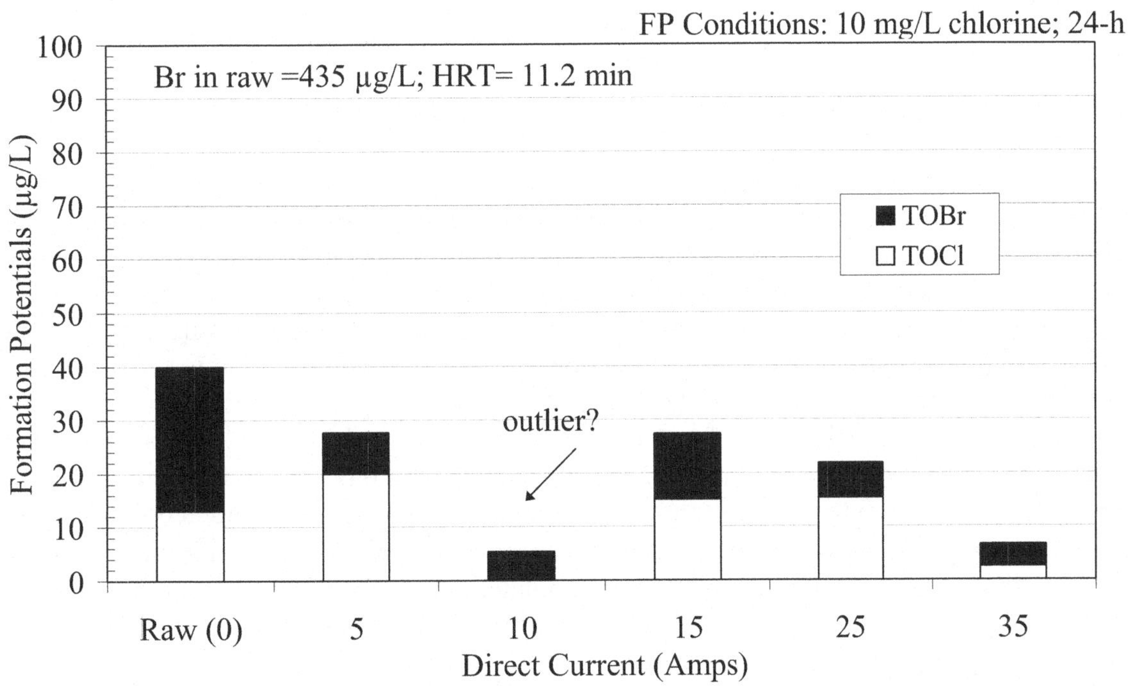

Figure 4.24 Effect of electrolysis on TOBr and TOCl formation potentials under ambient bromide conditions

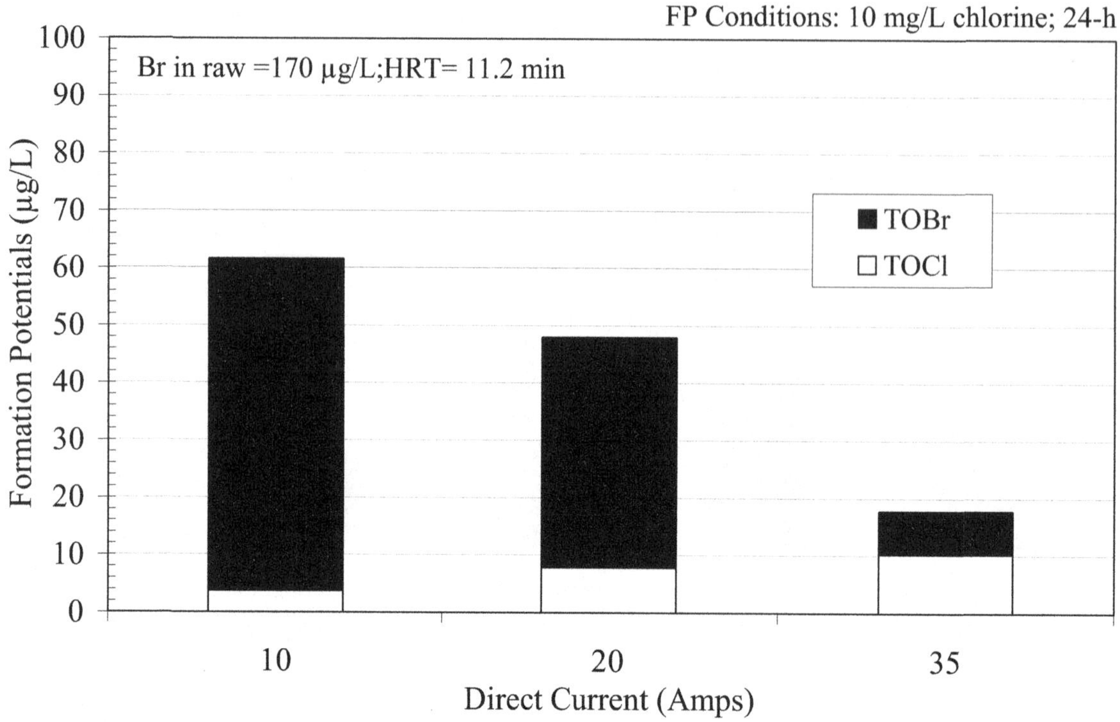

Figure 4.25 Effect of electrolysis on TOBr and TOCl formation potentials under spiked bromide conditions

REACTOR TESTING FOR MICROORGANISM INACTIVATION

Previous research has shown that electrolysis results in shifts in ORP and that the ORP shift is enough to suppress the growth of naturally occurring heterotrophic bacteria (Kimbrough et al. 2006). This is accomplished by a change in the ORP of the bacterial cell, which can reduce the potential and free energy. In these experiments, the ORP was changed without the addition of oxidizing or reducing agents. It is noted that the actual mechanism for suppressing growth is still under debate and the effect of ORP shift on microorganism growth suppression is just a theory.

To investigate this hypothesis, electrolyzed samples were quenched with sodium thiosulfate to reduce the halines in the water, and then raw and electrolyzed samples were autoclaved, inoculated with 1 mL of raw water, and plated each day for a period of 7 days. An 11.2-min HRT was used in the electrolytic reactor. Ambient raw water bromide conditions were used (i.e., 140 µg/L). ORPs were measured in all samples and summarized in Table 4.6. ORPs measured for other relevant experiments are also tabulated. A shift in ORP from a range of -13 to -30 mV to a range of -34 to -53 mV is observed as the current increases from 5 to 35 amps. The same trend was observed in all other experiments (results not shown).

The results are summarized in Figure 4.26 and further detailed in Figure 4.27 for the 7-day incubation. No effect on microbial growth was observed during the first two days. Days 2 and 3 show there is a growth stimulation in direct correlation to the applied current. Days 4, 5, and 6 show growth greatest in the middle ranges of the applied current. Day 7 shows a dramatic inhibition shifts at higher current.

Table 4.6
ORP shift (mV) measured during electrolysis

	Amperage							
Exp	Raw	5	10	15	20	25	30	35
1	-43.5	-29.3	-27	-31	-31.5	-36.5	-31	-40.5
2		-25	-34	-29	-30.9			-53
3		-13		-19.4	-27.5			-34.6
4		-13.2		-30.8	-33.3			-41
5		-27.3	-26.3	-32.6		-31		-40.3

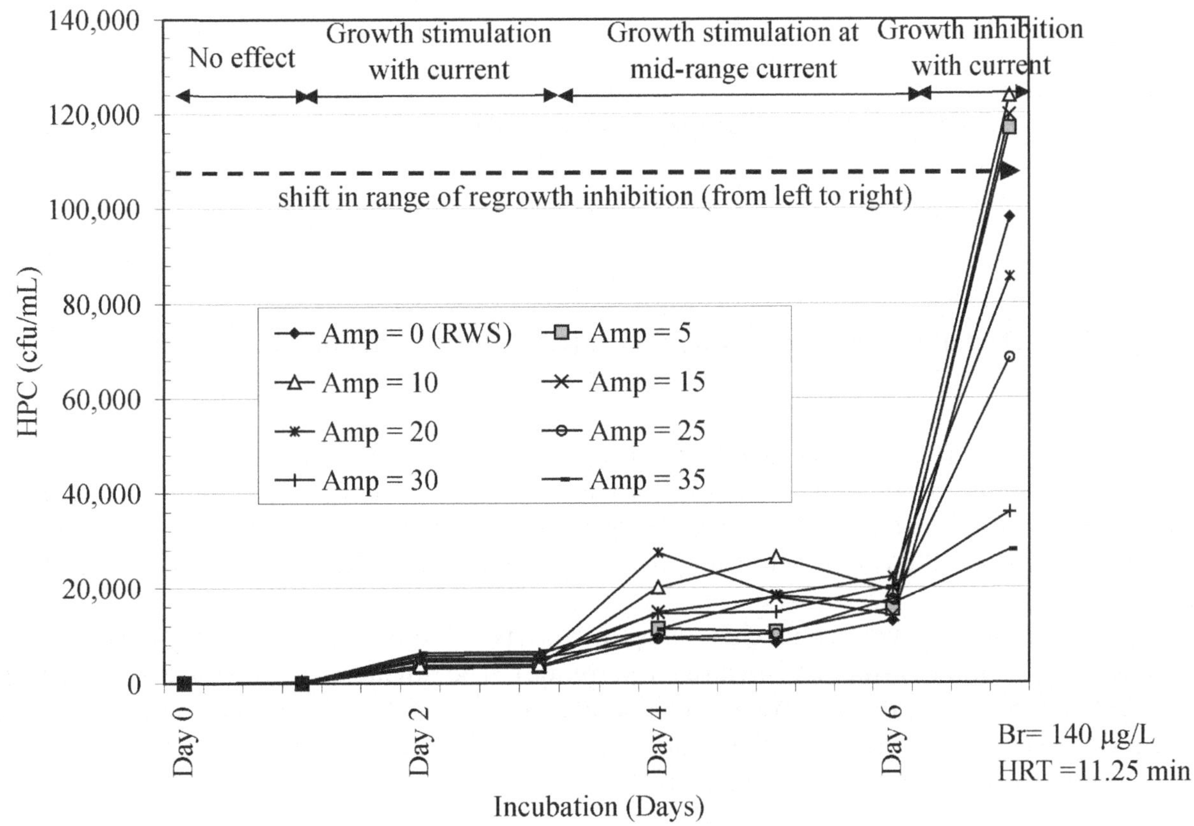

Figure 4.26 HPC in electrolyzed samples over 7 days of incubation

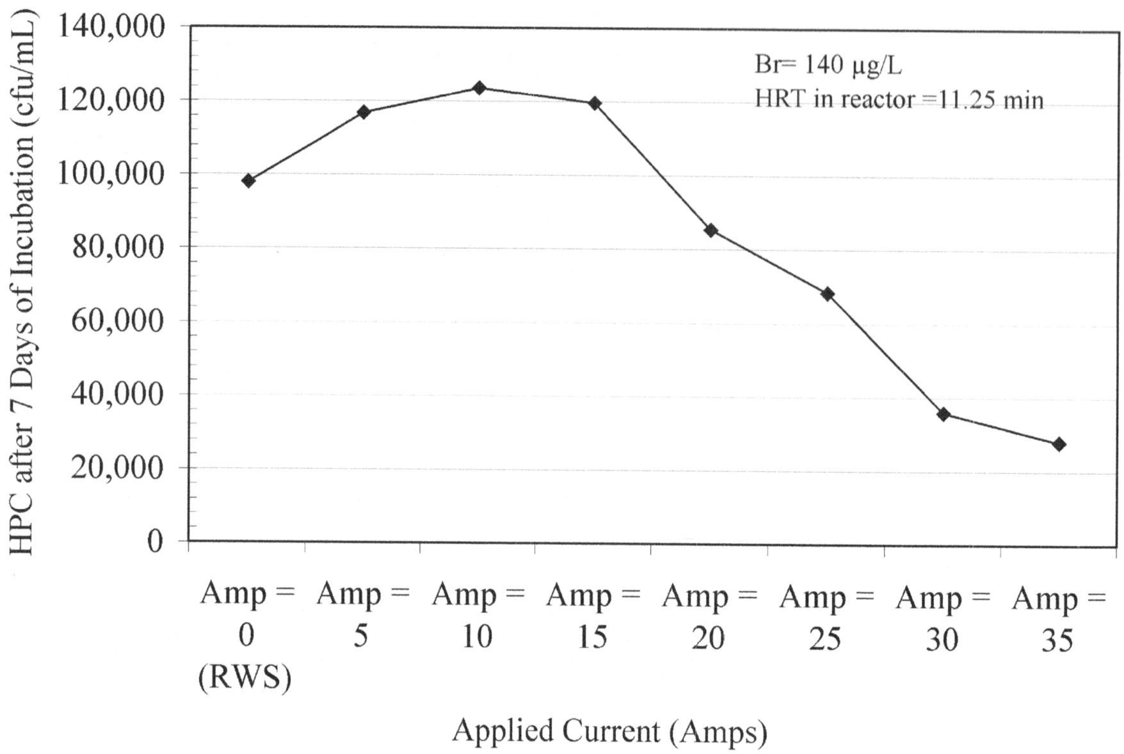

Figure 4.27 Effect of applied current on HPC after 7 days of incubation

The experiment was repeated by plating the samples every 3 days for 21 days. While growth stimulations are observed at the lower current ranges (5 to 10 amps), inhibitory growth effects are observed in the samples electrolyzed at higher currents after 9 days of incubation (Figure 4.28). After 18 days of incubation, the re-growth appear to be inhibited regardless of the applied current.

Limited testing was conducted using a variety of other dechlorinating agents (i.e., ascorbic acid, sodium sulfite, and hydrogen peroxide). The smallest re-growth was associated with the hydrogen peroxide. After two weeks of incubation, no HPC was detected in the samples, as opposed to 6.56×10^3 cfu/mL in the samples quenched with sodium thiosulfate.

More work is required to establish the likely cause of growth inhibition.

SIGNIFICANT FINDINGS OF REACTOR TESTING

The following summarizes the significant findings to date.

Reactor Hydraulics

A T_{10}/T ratio of 0.52 was measured in the electrolytic reactor. The ratio is indicative of average baffling conditions in a clearwell, with some short-circuiting.

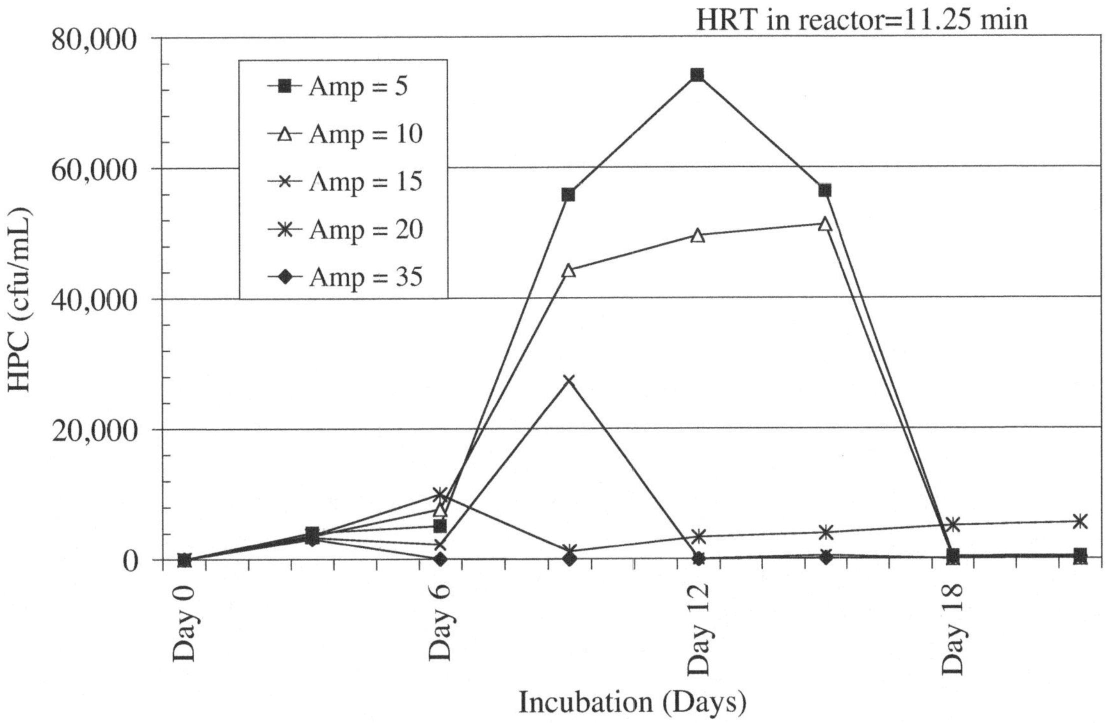

Figure 4.28 HPC regrowth in electrolyzed samples over 21 days of incubation

Raw Water Quality

Because the bromide levels in the raw water were close to their historical lows over the testing period, bromide was spiked in several of the experiments to a target of 400 to 500 μg/L. The chloride levels were slightly lower than historical ranges, at 45 to 47 mg/L, compared to a historic average of 59 mg/L. The TOC concentration was within historical ranges at approximately 2.5 mg/L. Even though this concentration is moderate, the organic matter in the raw water has been historically recalcitrant for removal by conventional treatment and requires high concentrations of coagulant.

Bromide Removal and Bromine Accumulation

The oxidation of bromide to bromine appeared to be a direct function of the applied current. This was observed regardless of the initial bromide concentration (150 to 435 μg/L). Even though there appeared to be no limit to the bromide oxidation to bromine, an accumulation of bromine was observed at higher applied current amperages. It appeared that below a threshold concentration of bromide in the water phase (i.e., 50 to 60 μg/L), the bromine volatilization was hindered.

The bromide oxidation appeared to be rate dependent, with very little oxidation of bromide to bromine occurring at the lower contact times (i.e., 3.2 min) and under the applied current amperages (up to 35 amps). However, as discussed in Chapter 3, mixing was demonstrated to negatively impact the bromide oxidation reaction. Thus, under the higher flowrates tested, sub-optimal conditions for bromide oxidation exist not only because of lower

HRTs, but also because of the occurrence of mixing. Optimal current ranges were determined for each hydraulic retention time condition and were defined in terms of both bromide removal and bromine accumulation minimization. A 15 amp current was optimal when the contact time was high (i.e., 11 minutes), whereas higher current amperages of 25 to >35 amps were required when the contact times decreased to 5.6 and 3.2 minutes.

When higher power (58 to 120 amps) was applied under low HRT (as low as 1.5 min) and high flow (up to 15 gpm), bromide was oxidized to bromine. Bromine accumulation was once again observed, limiting the overall bromide removals. A 1.5-min HRT under high turbulence conditions would require a power range of 2.4 to 5.8 kW to achieve maximum bromide removal and minimal bromine accumulation.

The overall bromide removals (μg/L per amp of applied current) were significantly higher at the 11-min HRT as compared to 5.6-min HRT and lower. The process is limited by the accumulated bromine fraction in the water.

Limited data from depth sampling have shown that the majority of the bromide removal was occurring in the first 10 cm of DSA plate depth under applied current and flowrate conditions of 35 amps and 2 gpm, respectively. Beyond that depth, bromine accumulation appears to be taking place. The implications can be extremely important. The required HRT would drop from 11 minutes to 1.2 minutes, which would significantly decrease the required footprint, volume requirements, and required number of DSA plates. In addition, bromine accumulation would be minimized, which could decrease the DBPFPs.

Despite the fact that sub-optimal bromide removal was measured when 15 gpm current was used (1.5-min HRT), the occurrence of mixing and turbulence within the reaction zone under that flow condition may have negatively impacted the bromide oxidation reaction, as discussed in Chapter 3 (mechanistic effects). Thus an HRT of 1 to 1.5 min measured under low turbulence conditions created a better environment for bromide oxidation as opposed to the same HRT under high flow/turbulence conditions.

DBP Formation Potential

In parallel to the bromide and bromine findings, DBPFPs appeared to decrease only under optimal current/power conditions (i.e., conditions associated with high bromide removals and low bromine accumulation). Applying too high of a current increased the formation potentials to levels higher than those in the raw water samples. An increase of chlorinated DBP species was observed in certain cases, likely due to the formation of chlorine as part of the electrolysis process. Brominated DBP species appeared to decrease in general. Under the high bromide spike conditions, the bromine incorporation factor (n) in THM4 and HAA6 was observed to decrease when an optimal current was applied. Although the data are limited, optimal ranges of current for decrease of DBPFP were equivalent to those for bromide removal and bromine minimization.

Under optimal current conditions, the removal of THMFP was approximately five times greater in samples electrolyzed under the 11-min HRT as compared to 5.6-min HRT and lower.

Dehalogenating the electrolyzed samples resulted in a small decrease in THMFP (approximately 20 μg/L). Thus, enhancing the natural stripping of accumulated bromine would be beneficial. The use of hydrogen peroxide as a quenching agent should be tested.

Speciated TOX (TOCl and TOBr) formation potentials were also observed to decrease in electrolyzed samples. However, unlike the THM, and HAA formation potentials, these halogenated species appeared to continue to decrease with higher current. More research in this matter is needed to confirm the findings.

Microbial Growth

An inhibitory effect was observed on growth of heterotrophic bacteria as a direct effect from electrolysis and ORP shift. The effect was more intense over longer periods of incubation (i.e., greater than 9 days). Limited data with hydrogen peroxide as the quenching agent showed that the bacterial regrowth is completely inhibited after 1 to 2 weeks of incubation. More research is needed to confirm these findings and to establish the likely cause of microbial regrowth inhibition.

CHAPTER 5
ASSESSMENT OF SYNERGY BETWEEN ELECTROLYSIS AND CONVENTIONAL TREATMENT

This chapter summarizes the findings from the assessment of the implementation of electrolysis within a conventional WTP. The discussion focuses on any synergistic effect between electrolysis and downstream conventional treatment processes. Data presented compare key parameters measured in waters that have undergone electrolysis alone to waters that have undergone electrolysis and conventional treatment. The chapter is structured under four main sections: electrolysis+CFS, electrolysis+ozonation, electrolysis+ozonation+CFS, and electrolysis+CFS+ozonation.

ELECTROLYTIC REACTOR AND DOWNSTREAM COAGULATION/ FLOCCULATION/SEDIMENTATION

Because electrolysis removes a fraction of the bromide in the water, and as long as the process can be optimized to minimize free available bromine and chlorine, the electrolytic process would likely be placed upstream of ozonation, and chlorination/chloramination. Since preozonation is practiced at several water treatment plants, this would place the electrolytic reactor at the head of the treatment plant (Figure 5.1). This section presents the results of testing electrolysis upstream of conventional CFS, without ozonation.

Results and Discussion

Bromide Removal and Bromine Accumulation

Raw and electrolyzed samples were subjected to CFS to investigate the effect of electrolysis+CFS on bromide removal and bromine accumulation.

It should be noted that only samples electrolyzed with a 35-amp current and 2 gpm flowrate were subjected to CFS. As presented in Chapter 4, although this current achieved the highest bromide removals, it was not optimal in terms of bromine minimization and decrease of DBPFP. In hindsight, the CFS should have been conducted on samples electrolyzed at 15 amps of current to assess the true benefits. Nevertheless, the preliminary findings from this phase of testing were promising.

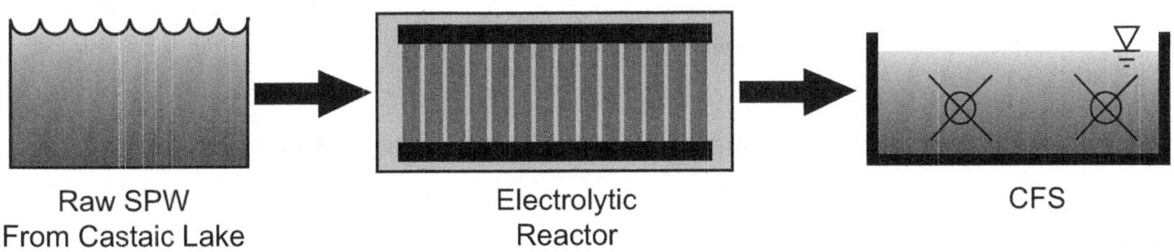

Figure 5.1 Schematic of electrolytic reactor upstream of CFS

The percent total bromide (i.e., sum of bromide anion and bromine) removal and fraction of bromine to total bromide are illustrated in Figure 5.2. Focusing on the electrolyzed sample, the application of 35 amps of current at 2-gpm flowrate resulted in up to 69 percent bromide removal.

As anticipated, CFS treatment of raw water did not result in any measurable bromide removal (i.e., 3 percent removal associated with the highest ferric chloride dose of 20 mg/L). However, when focusing on the effect of CFS on bromide electrolysis, 80 to 83 percent bromide removals were measured upon addition of 5 to 20 mg/L ferric chloride. Thus, it appears that CFS treatment had a synergistic effect on bromide electrolysis, increasing the bromide removals by 14 to 18 percent. On the other hand, the fraction of bromine in those samples appeared to increase as well, from 69 percent in the electrolyzed sample to about 90 percent in the CFS-treated electrolyzed sample. More data will be required to confirm these results.

DBP Formation Potential

As expected, CFS treatment, using 20 mg/L ferric chloride, decreased the THMFPs in both non-electrolyzed (from 192 μg/L to 36 μg/L) and electrolyzed samples (from 165 μg/L to 61 μg/L) (Figure 5.3). This is also represented by the decrease in TOC concentration of up to 50 percent as a result of adding 20 mg/L ferric chloride.

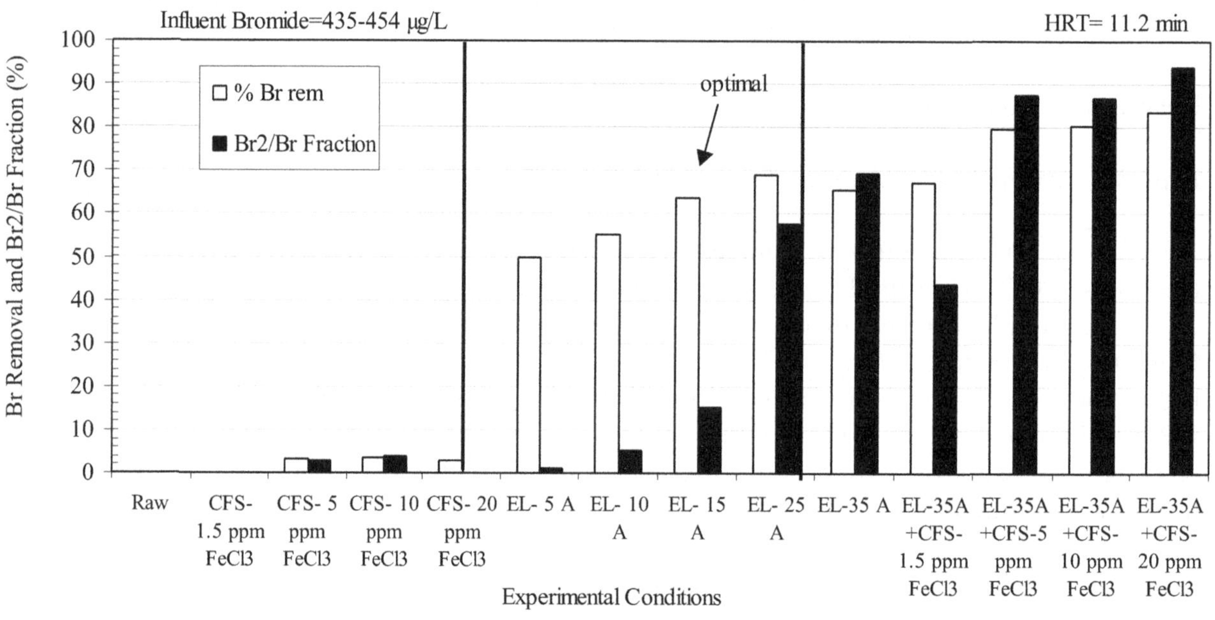

Figure 5.2 Effect of CFS treatment on bromide electrolysis (1.5 to 20 mg/L $FeCl_3$)

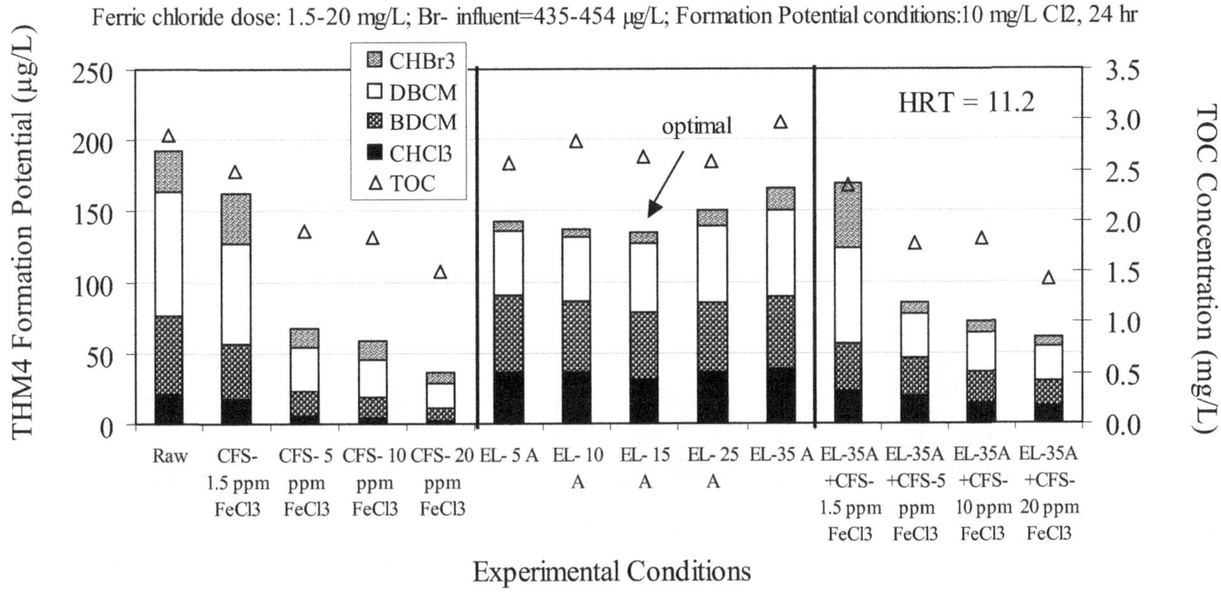

Figure 5.3 Effect of CFS treatment on THMFP in electrolyzed samples

The THMFP data in electrolyzed samples compliment the bromide/bromine data. The lowest THMFP was observed in the sample electrolyzed at 15 amps, which confirms results presented in Chapter 4. At this current condition, the THMFP in the electrolyzed sample was 135 µg/L, which was lower than 192 µg/L in the non-electrolyzed water sample. Focusing on speciated THMs, when comparing the non-electrolyzed sample to the sample electrolyzed at 15 amps, a drop in brominated THM species (bromoform, dibromochloromethane, and dichlorobromomethane) is observed, whereas chloroform appears to increase. The bromine incorporation factor (n) in THM4 decreased from 1.4 in the raw water sample to 0.95 in the sample electrolyzed with 10 to 15 amps of current.

When focusing on the combination of electrolysis+CFS, a synergistic effect is observed when ferric chloride was added at doses of 5 mg/L and higher. In the sample electrolyzed at 35 amps of current, the THMFP dropped from 165 µg/L (without CFS) to 61 µg/L when 20-mg/L ferric chloride were added. This is, again, a result of the drop in TOC concentration from 3.0 to 1.4 mg/L.

However, it appears at first glance that the THMFPs in the CFS-treated electrolyzed samples were higher than in the CFS-treated non-electrolyzed water. Again, this is likely caused by the fact that the combination of electrolysis + CFS was only tested in the sample electrolyzed at 35 amps. At this current level, the bromine to bromide was too high (69 percent), resulting in high THMFPs.

The experiment was repeated by applying a lower current of 10 amps and a ferric chloride dose of 1.5 mg/L (Figure 5.4). At this current level, 25 percent bromide removal was achieved with a minimal residual bromine to bromide fraction of 2 percent. When comparing CFS-treated non-electrolyzed water to CFS-treated electrolyzed water, a drop of 16 µg/L in THMFP was measured, mainly attributed to a drop in brominated THM species (bromoform, dibromochloromethane, and dichlorobromomethane).

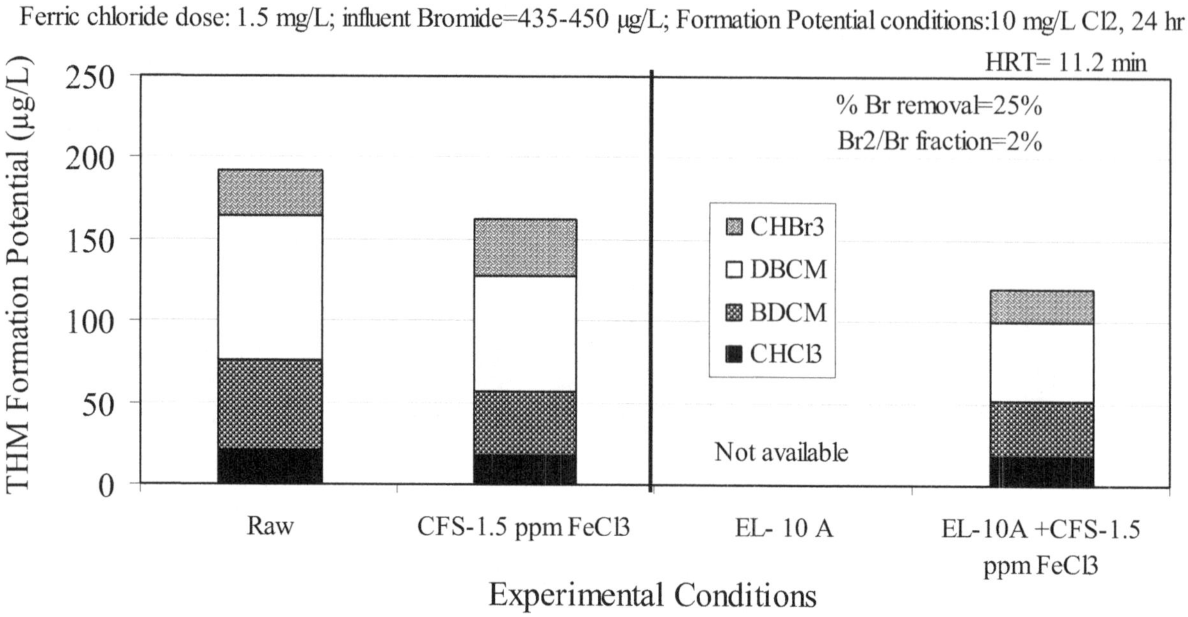

Figure 5.4 Effect of CFS treatment on THMFP under low current electrolysis conditions

Heterotrophic Bacteria Regrowth

As presented in Chapter 4, hindered microbial growth was observed in electrolyzed samples as compared to non-electrolyzed water samples. When CFS is added downstream of electrolysis, the same trend in hindered growth is observed. Regardless of the coagulation conditions, effluent samples electrolyzed at 35 amps measured significantly inhibited growth of heterotrophic bacteria as compared to CFS-treated non-electrolyzed water samples (Figure 5.5). Figure 5.6 illustrates the bacterial growth over time. While heterotrophic bacteria did not grow in the coagulated electrolyzed sample throughout the 21 days of incubation, HPC increased initially, then decreased over time in the CFS-treated non-electrolyzed sample. No significant coagulant effect was observed on the growth of heterotrophic bacteria.

INVESTIGATION OF SYNERGY BETWEEN ELECTROLYTIC REACTOR AND DOWNSTREAM OZONATION PROCESS

Since several water treatment plants, including CLWA, incorporate preozonation in their treatment scheme, the effect of integrating an electrolytic process upstream of ozonation was investigated in this portion of the study (Figure 5.7).

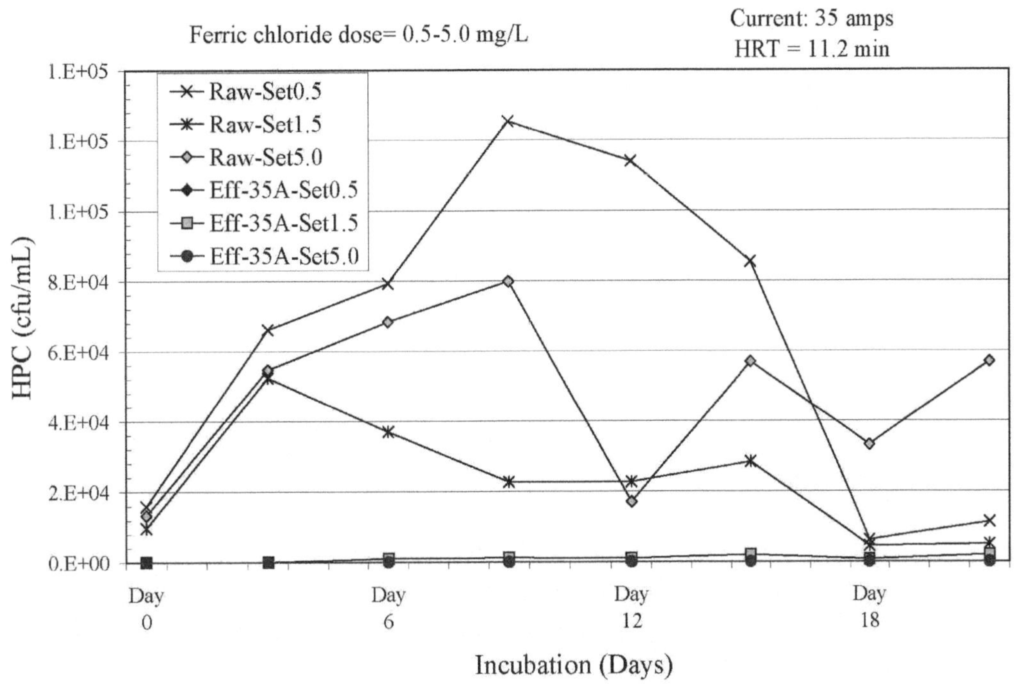

Figure 5.5 Effect of electrolysis and CFS on the evolution of HPC over 21 days

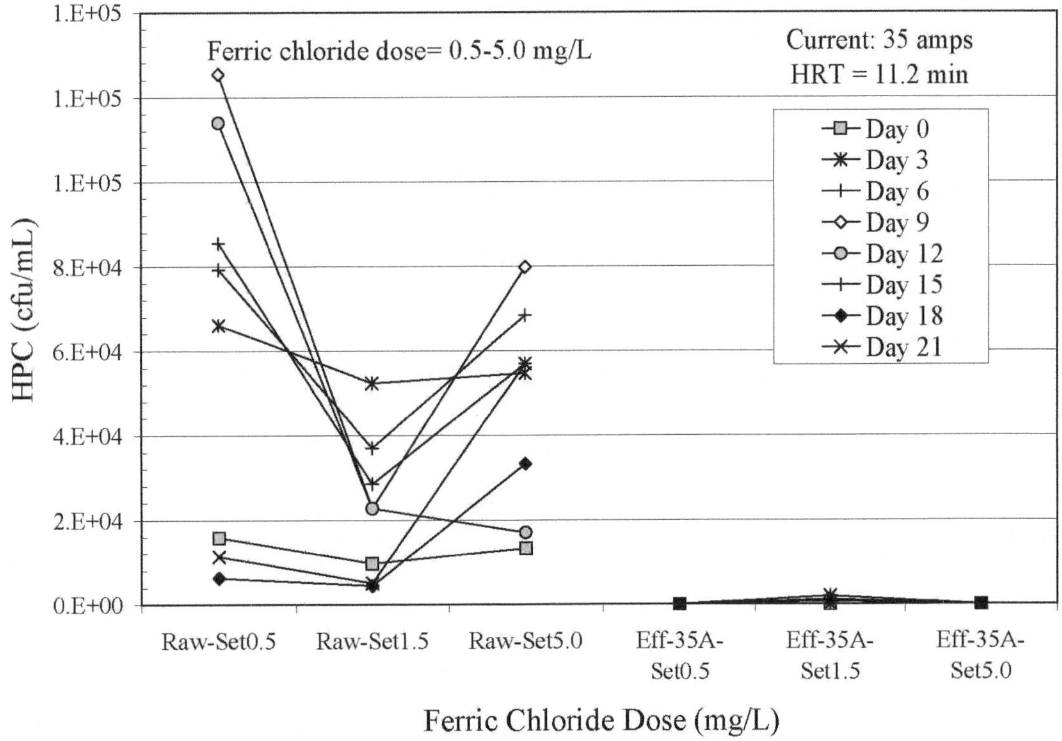

Figure 5.6 Effect of coagulant dose on HPC over time

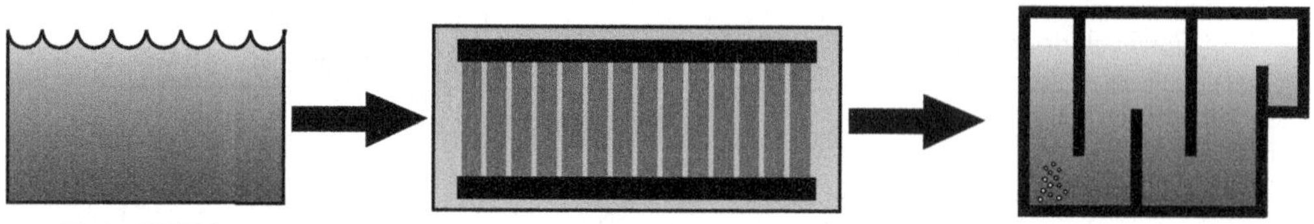

Figure 5.7 Schematic of electrolytic reactor upstream of ozonation

Results and Discussion

DBP Formation Potential

Raw and electrolyzed waters were subjected to ozonation. Two ozone conditions were tested as presented in Table 5.1. The electrolytic reactor ran under the following conditions: 11.2-min HRT and 35 amps (i.e., 945 watts). In those experiments, the bromide was spiked up to a concentration of 400 to 435 μg/L.

Due to the constraints of the experimental plan, limited data on electrolysis+ozone are available. More data can be found in the next sections on combined CFS+ozonation.

As anticipated, ozonation, which oxidizes the natural precursors of disinfection by-products, appeared to decrease the THMFP in the non-electrolyzed water from 192 μg/L to 150 μg/L (Figure 5.8). Unfortunately, no data were available in electrolyzed samples, but the same trend is anticipated. Additional experiments are required to confirm these findings. More data on THMFPs are available in the next section on combined CFS+ozonation.

Table 5.1
Ozonation conditions

Conditions 1	
Numbers of columns	1
HRT	5 min
Transferred ozone dose	0.6 mg/L
Ozone residual in column 1	0.3 mg/L
Conditions 2	
Numbers of columns	2
HRT	8 min
Transferred ozone dose	0.4 mg/L
Ozone residual in column 2	0.07 mg/L

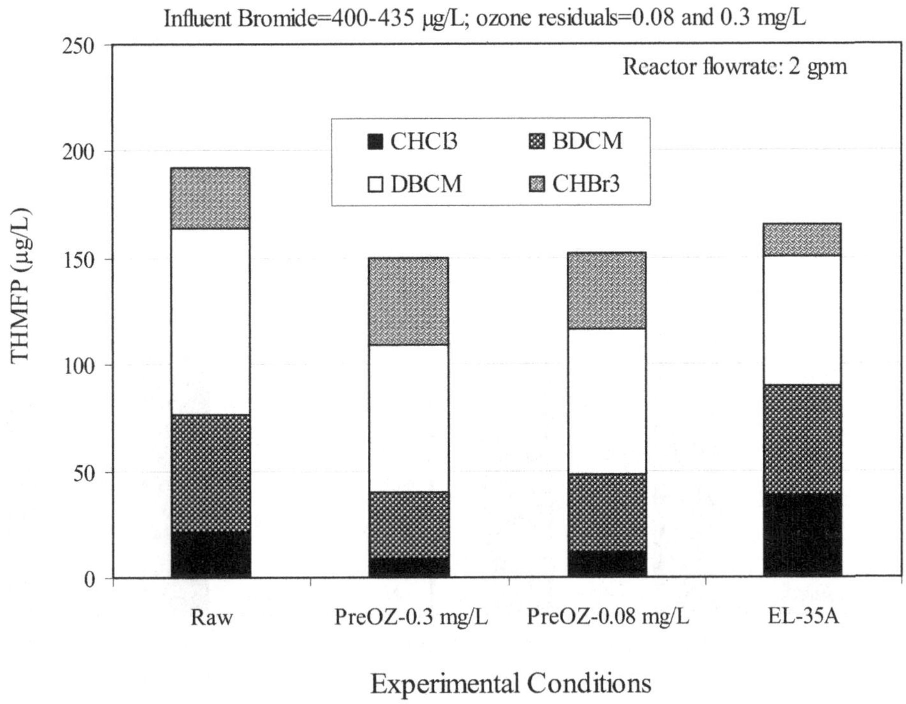

Figure 5.8 Impact of ozonation on THMFP

Bromate Formation

The bromate formed upon ozonation in raw and electrolyzed samples is illustrated in Figure 5.9. Since a portion of the bromide was removed by electrolysis, a small but consistent decrease in the bromate formation potential was observed in electrolyzed samples. Under identical conditions of ozonation, a 1.1- to 1.8-μg/L decrease in bromate concentration was observed when comparing non-electrolyzed and electrolyzed samples. It is noted that no bromate was detected in ozonated non-electrolyzed water samples when the bromide concentration was 150 μg/L (ambient).

Additional batch ozone experiments were conducted during which ozone was dosed to raw and electrolyzed waters such that instantaneous residuals of 0.5 mg/L, 1.5 mg/L, and 3 mg/L were measured. The instantaneous ozone demands ranged from 0.5 mg/L to 1.53 mg/L. In addition, initial bromide was spiked from 130 μg/L up to three target concentrations: 220 μg/L, 270 μg/L, and 400 μg/L. In these experiments, residual chlorine/bromine in electrolyzed samples was quenched by adding stochiometric amounts of hydrogen peroxide to minimize ozone demand and other reactions of bromine/chlorine with the organic matter. The electrolytic reactor was operated under an 11.2-min HRT, and the direct current ranged from 15 to 35 amps. The ozonated raw and electrolyzed samples were allowed to react for 4 days before samples were collected for bromate analyses. Results are illustrated in Figures 5.10 and 5.11. The results show no consistent change in bromate with applied amperage. The optimal current for decrease of bromate formation potential varied in each sample. In addition, the beneficial decrease in bromate concentration was more pronounced when the high ozone residual was used. Due to data gaps with the low ozone residual test, results have been excluded. More testing will be required to confirm these results.

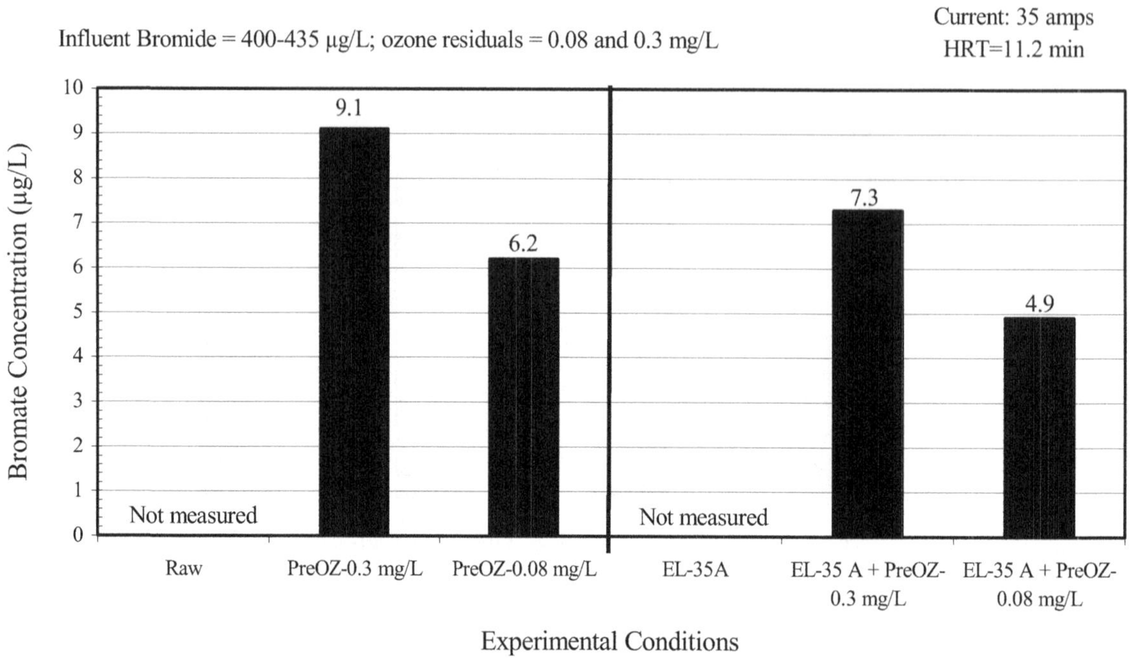

Figure 5.9 Impact of electrolysis and continuous flow ozonation on bromate formation

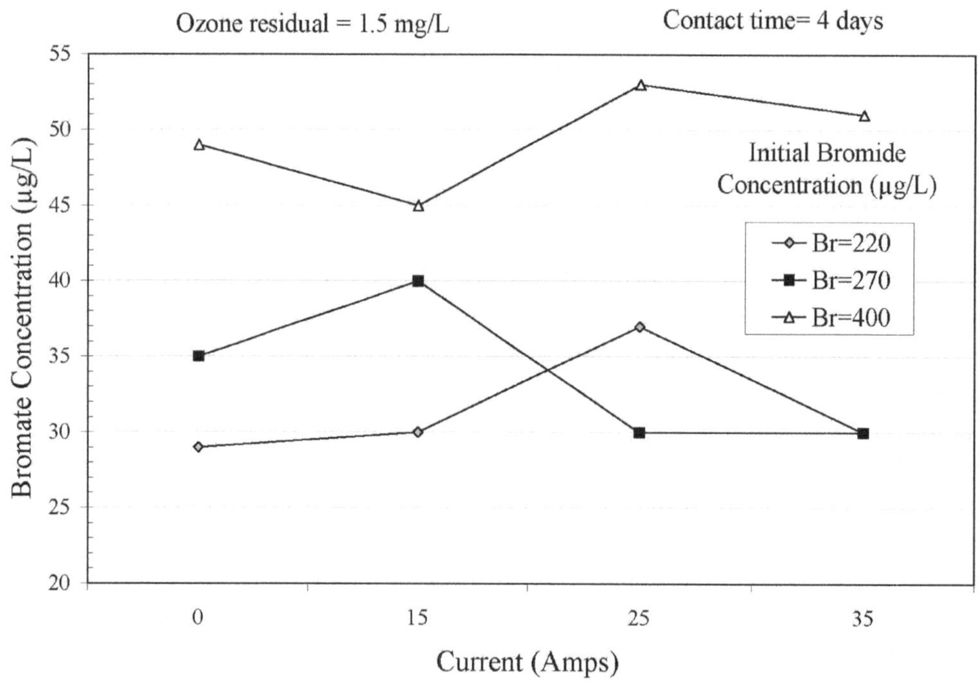

Figure 5.10 Impact of electrolysis and moderate ozone dose on bromate formation

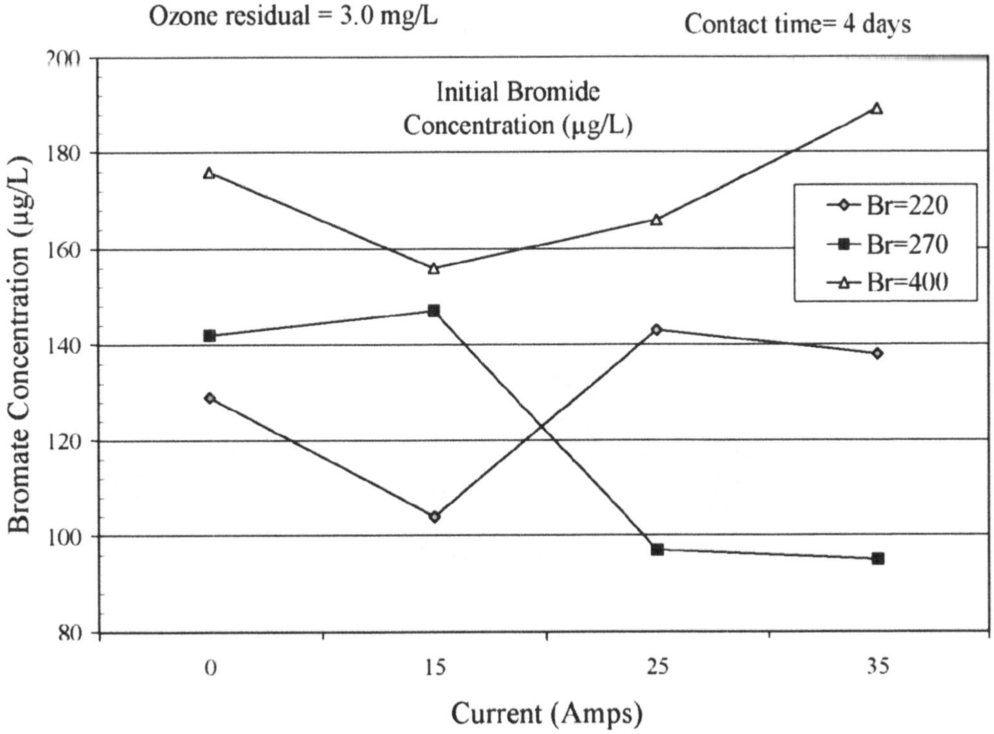

Figure 5.11 Impact of electrolysis and high ozone dose on bromate formation

ELECTROLYTIC REACTOR, OZONATION, AND CFS

Electrolysis upstream of ozonation and CFS was mimicked in these sets of experiments (Figure 5.12).

Results and Discussion

Bromide Removal and Bromine Formation

When comparing electrolyzed samples to samples subjected to electrolysis in addition to ozonation and CFS, the percent bromide removal remains more or less unchanged while a decrease in the residual bromine was observed (Figure 5.13). A decrease from 77 percent to 50 percent in the bromine fraction was observed upon ozonation. The fraction dropped further to 10 percent when CFS treatment was applied following ozonation. The residual bromine may have been stripped during the ozonation and jar mixing conditions.

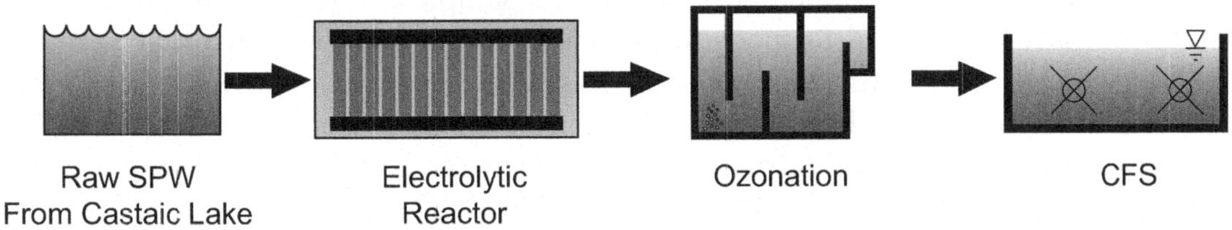

Figure 5.12 Schematic of electrolytic reactor upstream of ozonation+CFS

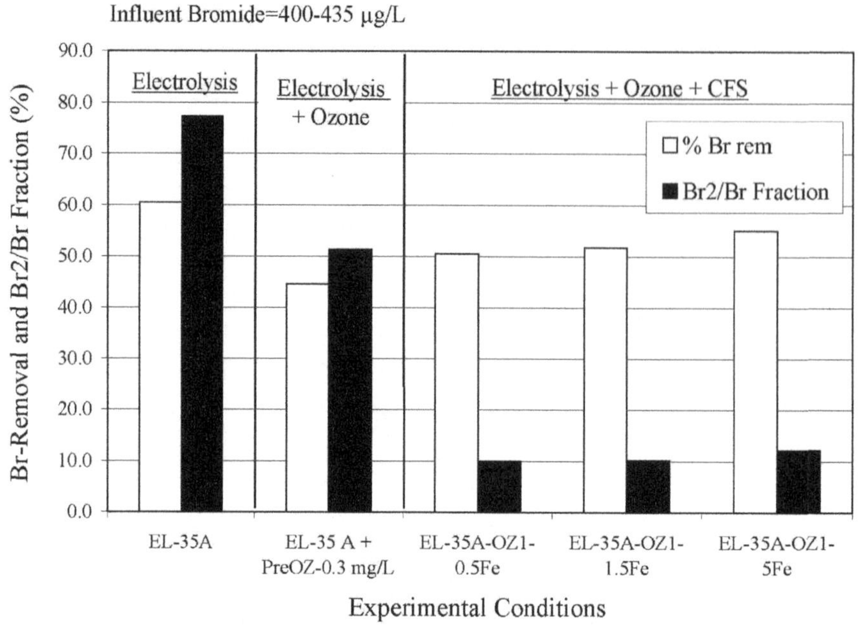

Electrolytic reactor operated at 35 amps and 11.2 min HRT; ferric chloride dose range of 0.5-5 mg/L; ozone residual of 0.3 mg/L and 5 min HRT

Figure 5.13 Effect of combined ozonation+CFS on bromide oxidation

DBP Formation Potentials

Ozonation, and CFS appeared to benefit the electrolysis process in terms of decreasing the THMFP from 165 μg/L to 110 μg/L when a 5-mg/L ferric chloride dose was applied (Figure 5.14). These results compliment the findings of electrolysis+CFS, showing that a 5 mg/L ferric chloride dose and higher appeared to compliment the electrolysis process.

TOCl and TOBr Formation Potentials

Select samples were subjected to speciated TOXFP testing. As expected, CFS and the combination of preozonation+CFS appeared to significantly decrease the TOBr and TOCl formation potentials from a combined total (TOBr+TOCl) of 40 μg/L in the raw water to 3 μg/L with preozone+CFS (Figure 5.15).

Inconsistent results were obtained in electrolyzed samples. While the application of increasing levels of current appeared to increase the TOBr formation potential in CFS-treated samples, applying up to 35 amps appeared to decrease the formation potentials when preozonation+CFS were mimicked. These results differ from those presented in Chapter 4 that show that applying increasing levels of current resulted in decreasing TOBr formation potentials. More research on this topic would be appropriate (Figure5.16).

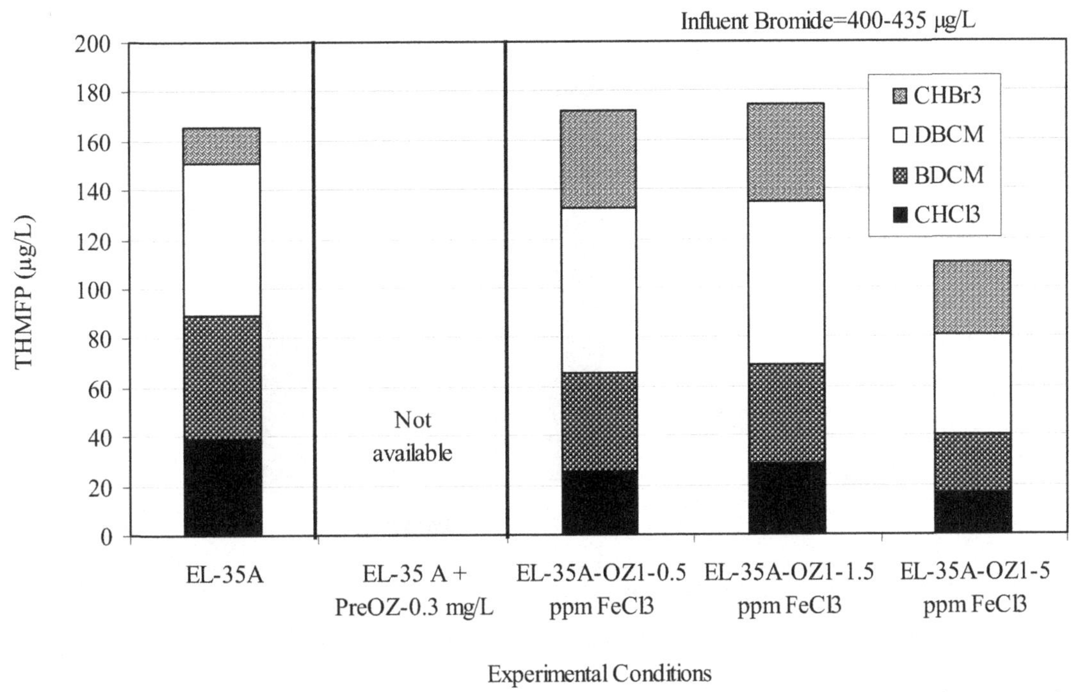

Electrolytic reactor operated at 35 amps and 11.2 min HRT; ferric chloride dose range of 0.5-5 mg/L; ozone residual of 0.3 mg/L and 5 min HRT

Figure 5.14 Effect of combined ozonation+CFS on THMFP in electrolyzed samples

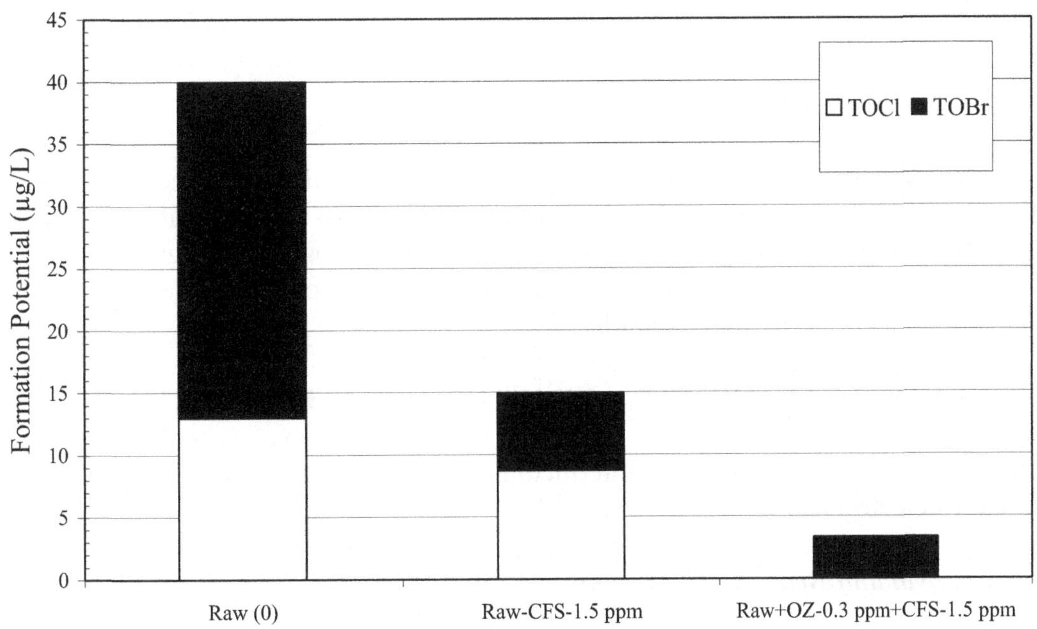

Figure 5.15 Effect of combined ozonation+CFS on TOBr and TOCl formation potentials in raw water

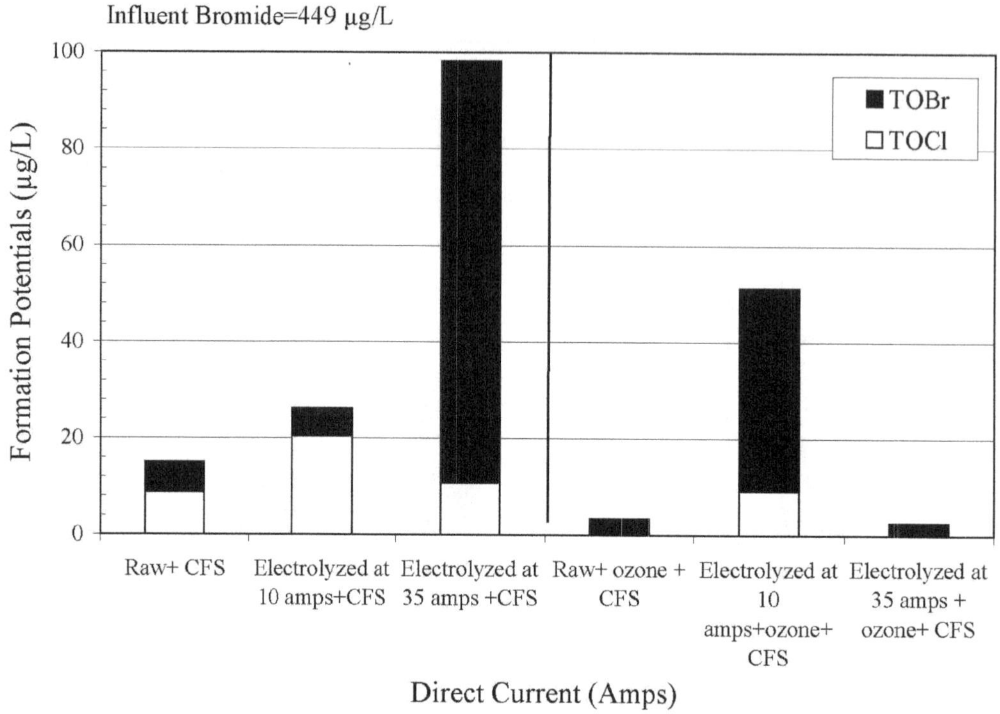

Formation potential conditions" 10 mg/L chlorine, 24 hrs. Electrolytic reactor conditions: 10 and 35 amps at 11.2-min HRT; CFS conditions: ferric chloride dose – 1.5 mg/L; Ozonation: 0.3 mg/L residual, 5-min HRT

Figure 5.16 Effect of combined ozonation+CFS on TOBr and TOCl formation potentials in electrolyzed water

Heterotrophic Bacteria Regrowth

An increase in the bacterial growth was observed in the ozonated electrolyzed water samples. Ozone has been documented to increase the fraction of biodegradable organic matter in natural waters, which is a food source for the heterotrophic bacteria (Figure 5.17). The HPC appeared to decrease over time. Although results are not included, the same trend was observed in ozonated non-electrolyzed water samples.

ELECTROLYTIC REACTOR, CFS, AND INTERMEDIATE OZONATION

Since many plants incorporate intermediate ozonation, this treatment scheme was tested downstream of electrolysis (Figure 5.18).

Bromide Removal and Bromine Formation

When comparing electrolyzed samples to samples subjected to additional CFS and CFS+ozonation, it appears that the residual bromine is being stripped by the jar testing mixing (bromine fraction decreased from 69 percent to 44 percent), and by the ozonation (bromine fraction further dropped to 13 percent) (Figure 5.19). No effect is observed on the bromide concentration. The same trend was observed when 10 amps of current were applied.

Results and Discussion

THM Formation Potentials

The addition of intermediate ozonation downstream of CFS appeared to decrease somewhat the DBPFP from 170 µg/L to 150 µg/L (Figure 5.20). It should be kept in mind that the applied ferric chloride dose in these experiments was 1.5 mg/L. A higher dose would likely further decrease the THMFP, as observed in the CFS experiments.

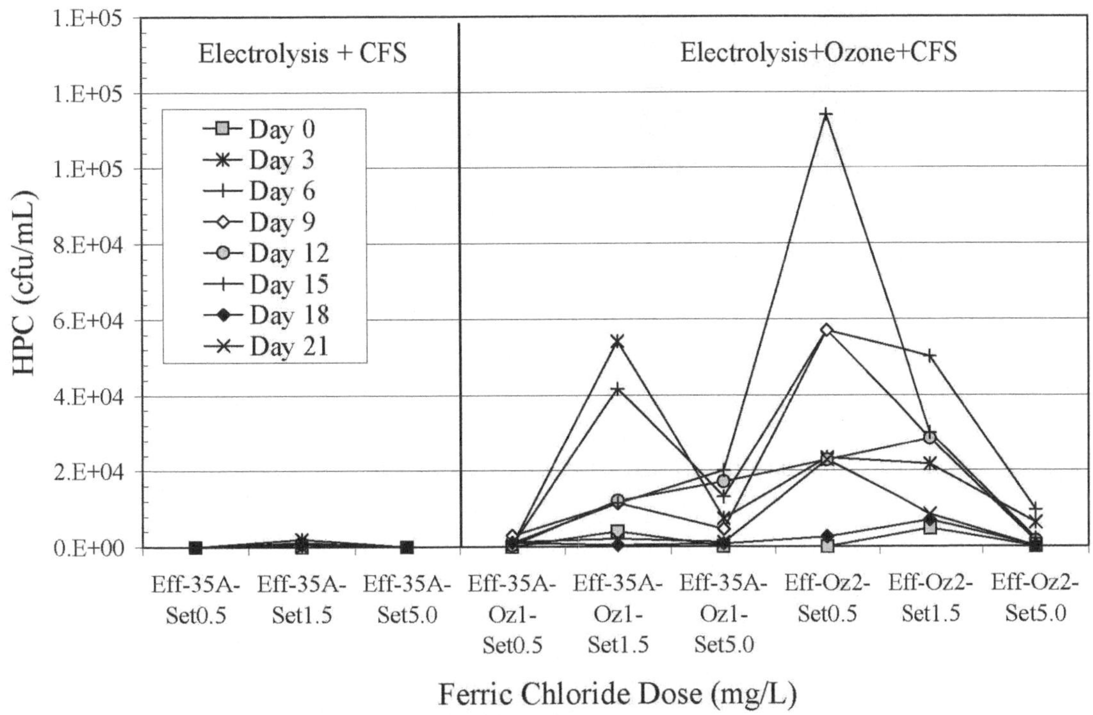

Electrolytic reactor conditions: 35 amps current, 11.2-min HRT; CFS conditions: 0.5- 5 mg/L ferric chloride; Ozone 1 conditions: 0.2 mg/L residual, 5 min HRT; Ozone 2 conditions: 0.07 mg/L residual, 10-min HRT.

Figure 5.17 Effect of ozonation on heterotrophic bacteria regrowth in electrolyzed waters

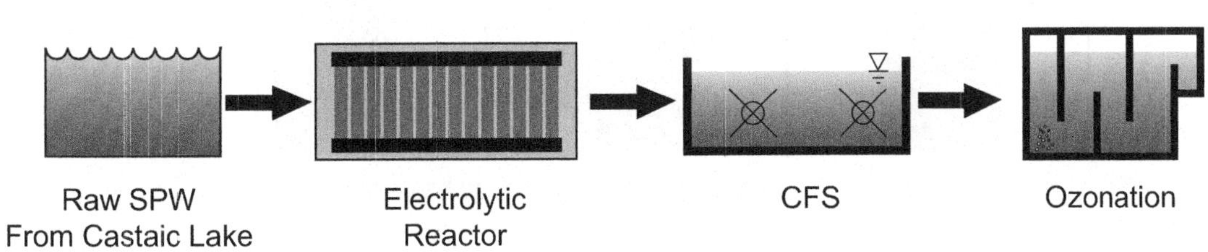

Figure 5.18 Schematic of electrolytic reactor upstream of CFS and intermediate ozonation

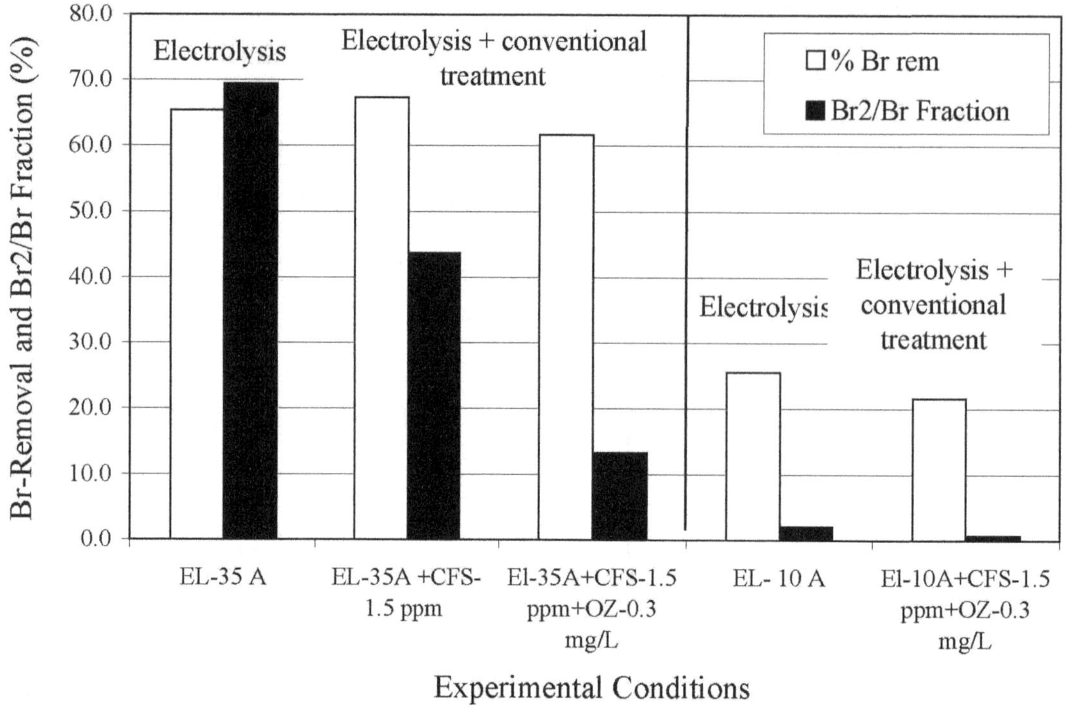

Electrolytic reactor conditions: 35 amps curremt at 11.2-min HRT; Ozone conditions: 0.2 mg/L residual, 5 min HRT; CFS conditions: 1.5-20 mg/L ferric chloride

Figure 5.19 Effect of CFS+ozonation on bromide oxidation

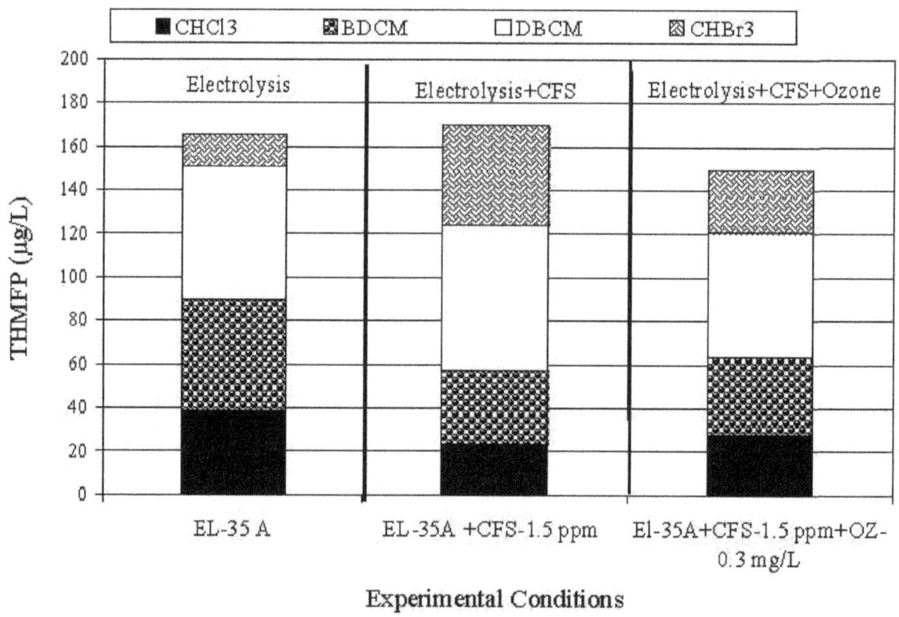

Electrolytic Reactor conditions: 35 amps current and 11.2-min HRT; Ozone conditions: 0.39 mg/L residual and 5-min HRT; CFS conditions: 1.5-20 mg/L ferric chloride

Figure 5.20 Effect of CFS+ozonation on THMFP in electrolyzed samples

Bromate Formation

Under identical conditions of ozonation and/or CFS, the bromate concentrations in the treated un-electrolyzed waters were 2 to 10 µg/L higher than those in the treated electrolyzed samples (Figure 5.21). Ozonated/un-electrolyzed water bromate concentrations ranged from 6 to 17 µg/L, whereas ozonated/electrolyzed water bromate concentrations were in the 4 to 7 µg/L range. This finding can probably be explained by the removal of the bromide ion from the raw water as a result of oxidation to bromine and volatilization during electrolysis.

EFFECT OF ELECTROLYSIS AND CONVENTIONAL TREATMENT ON FLAVOR PROFILE

Select raw and electrolyzed water samples were sent in for Flavor Profile Analysis (Table 5.2). Raw water from the Castaic Lake was characterized as grassy/fishy. The addition of conventional treatment gave the water an earthy/musty and fishy T&O characteristic. Electrolyzed treated and untreated waters were characterized as having a calcium scale odor (typical of water in a hot water kettle), chlorine, and earthy musty T&O characteristic. The chlorine is generated through the electrolysis process, and the calcium scale odor is likely due to hardness deposits within the interior of the reactor and DSA plates (Figure 5.22).

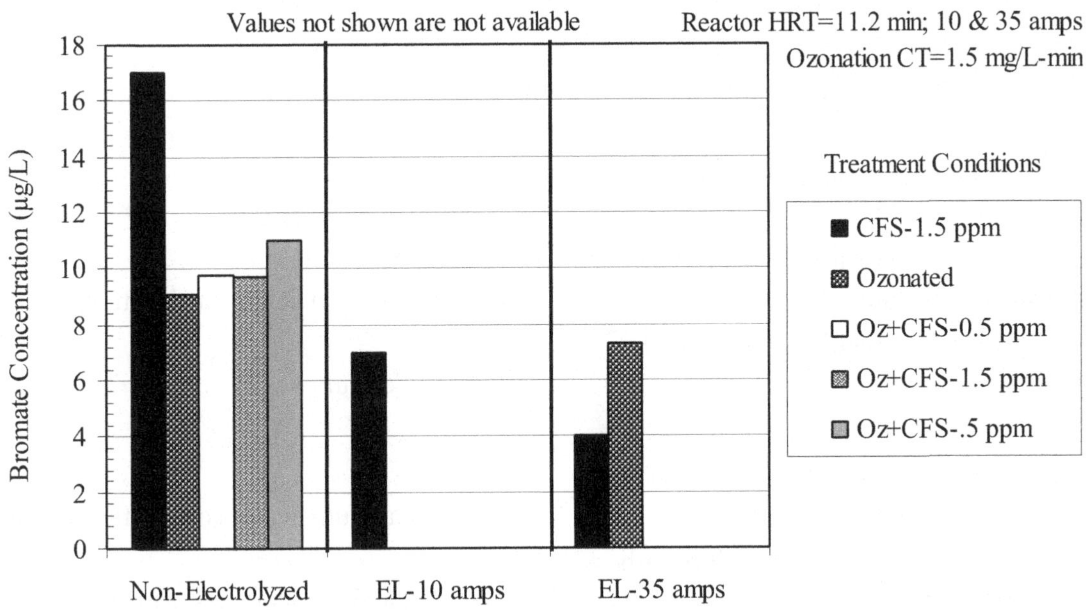

Electrolytic reactor conditions: 10 & 35 amps current, 11.2-min HRT; CFS conditions: 0.5-5 mg/L ferric chloride; Ozone conditions: 0.07 and 0.3 mg/L residuals

Figure 5.21 Effect of CFS+ozonation on bromate formation in raw and electrolyzed samples

Table 5.2
Taste and odor characteristics of raw and electrolyzed waters

Sample description	Current (Amps)	Flavor profile description
Raw	0	Grassy, fishy
Raw + 0.3 ppm ozone residual	0	Earthy/musty
Eff+ 0.3 ppm ozone residual	35	Calcium scale
Raw + CFS with 0.5 ppm ferric chloride	0	Fishy
Raw + CFS with 1.5 ppm ferric chloride	0	Earthy/musty
Raw + CFS with 5 ppm ferric chloride	0	Earthy/musty, fishy
Eff+ CFS-1.5 ppm	10	Ca scale, earthy/musty
Eff + CFS-0.5 ppm	35	Ca scale, earthy/musty, chlorine
Eff + CFS-1.5 ppm	35	Ca scale, fishy, alcohol, chlorine
Raw + CFS with 1.5 ppm ferric chloride + 0.3 ppm ozone residual	0	Earthy/musty
Eff + CFS with 1.5 ppm ferric chloride + 0.3 ppm ozone residual	10	Calcium scale
Eff + CFS with 1.5 ppm ferric chloride + 0.3 ppm ozone residual	35	Calcium scale
Eff + 0.3 ppm ozone residual + CFS with 0.5 ppm ferric chloride	35	Calcium scale
Eff + 0.3 ppm ozone residual + CFS with 1.5 ppm ferric chloride	35	Calcium scale, earthy/musty, chlorine
Eff + 0.3 ppm ozone residual + CFS with 5 ppm ferric chloride	35	Calcium scale
Eff + 0.07 ppm ozone residual + CFS with 0.5 ppm ferric chloride	35	Calcium scale, earthy/musty, chlorine
Eff + 0.07 ppm ozone residual + CFS with 1.5 ppm ferric chloride	35	Calcium scale, chlorine
Eff + 0.07 ppm ozone residual + CFS with 5 ppm ferric chloride	35	Calcium scale

Figure 5.22 Hardness scale on DSA surface

CONCLUSIONS OF SYNERGY TESTING

Four treatment configurations were tested during the course of the study to investigate whether there are any synergistic effects between electrolysis and conventional treatment:

Electrolysis + CFS
Electrolysis + Ozonation
Electrolysis + ozonation + CFS
Electrolysis + CFS + Intermediate ozonation

Results are summarized in Table 5.3.

Bromide Removal and Bromine Formation

Some synergistic effects were observed in terms of bromide removal when CFS was applied downstream of electrolysis. The addition of 5 to 20-mg/L ferric chloride increased bromide removals by 14 to 18 percent. However, the fraction of bromine in those samples appeared to increase as well, from 69 percent in the electrolyzed sample to about 90 percent in the CFS-treated electrolyzed sample.

When ozonation and CFS were mimicked downstream of electrolysis, a significant decrease in the fraction of residual bromine was observed (from approximately 70 percent in electrolyzed samples to 10 percent after ozonation+CFS), potentially due to stripping experienced during ozonation and jar mixing conditions. This could in turn lower DBPFPs.

Table 5.3
Combined effect of electrolysis and conventional treatment on water quality

Treatment	Bromide	Bromine	DBPFP	HPC	Bromate
Electrolysis + CFS	Increased removals by 14-18%	Increased bromine fraction from 69 to 90%	Decrease in THMFP from 165 to 86 μg/L with ≥ 5 mg/L ferric chloride	Hindered growth	Not applicable
Electrolysis + Ozonation		Stripping of bromine fraction from 77% to 50%	Drop from 192 to 150 μg/L in non-electrolyzed samples. Same trend anticipated in electrolyzed samples	Not tested	Decrease in bromate by 1 to 2 μg/L
Electrolysis + Preozonation + CFS		Further stripping of bromine fraction down to 10%	Decrease in THMFP from 165 to 110 μg/L upon addition of 5 mg/L ferric chloride	Initial increase in HPC, followed by decrease over time	Not tested
Electrolysis + CFS + Intermediate ozonation		Stripping of bromine fraction from 70 to 40% with CFS and further down to 13% with ozonation	Decrease in THMFP from 170 to 150 μg/L when ozone added downstream of CFS	Not tested	Decrease in bromate by 2 to 10 μg/L

Halogenated DBP Formation Potentials

In terms of halogenated DBPFPs, CFS is a well proven treatment process for removing DBP precursors. This was shown in the drop of THMFP in both the CFS-treated un-electrolyzed (from 192 to 36 μg/L) and electrolyzed water samples (from 165 to 61 μg/L). What was interesting however, is the observed synergistic effect of CFS on electrolyzed samples. At 10 amps of current, the THMFPs in the electrolyzed/CFS-treated samples dropped to levels lower than those in the CFS-treated un-electrolyzed samples.

Lower THMFPs (decrease from 192 to 150 μg/L) were measured in the non-electrolyzed sample when 0.08 to 0.3 mg/L ozone residual was mimicked, as opposed to without. The same trend is anticipated in electrolyzed samples.

A decrease in THMFP in electrolyzed samples from 165 to 110 μg/L was measured when ozonation +CFS (at 5 mg/L ferric chloride) were mimicked.

The addition of intermediate ozonation downstream of CFS appeared to somewhat decrease the DBPFP, from 170 to 150 μg/L.

Bromate Formation

A slight effect was observed when ozonation was preceded with electrolysis. A 1.1 to 1.8 μg/L decrease in the bromate formation potential was observed in electrolyzed samples as compared to un-electrolyzed water samples. This decrease is however small and within the experimental/analytical error range.

The addition of intermediate ozonation downstream of CFS appeared to lower bromate concentrations. Under identical conditions of ozonation and/or CFS, the bromate concentrations in the treated un-electrolyzed waters were 2 to 10 μg/L higher than those in the treated electrolyzed.

Microbial Growth

As presented in Chapter 4, when CFS is added downstream of electrolysis, the same trend in hindered heterotrophic bacteria regrowth is observed in the electrolyzed samples. Regardless of the coagulation conditions, effluent samples electrolyzed at 35 amps measured significantly inhibited growth of heterotrophic bacteria as compared to non-electrolyzed water samples

One potential detrimental effect is the observed increase in the heterotrophic bacteria growth in the ozonated un-electrolyzed and electrolyzed water samples. However, HPC appeared to decrease over time in both non-electrolyzed and electrolyzed ozonated samples.

CHAPTER 6
CONCEPTUAL SCALE UP ISSUES

CONCEPTUAL SCALE UP CONSIDERATIONS

Conceptual Design of Electrolytic Reactor

Pilot-scale data were used to investigate the scalability of the electrochemical reactor to a full-scale system. Key physical pilot data included the anode plate surface area, applied current, and flowrate. The optimum current level and minimum acceptable HRT condition were based on simultaneously achieving maximum bromide removal, minimal bromine accumulation, and minimum DBPFPs.

Pilot testing of the electrochemical reactor led to the following key findings relevant to scalability:

- The reactions that lead to bromide oxidation and volatilization may be occurring within a few millimeters of the anode plate surface. As such, the ratio of anode surface area to flowrate is an important design parameter.
- A high ratio of anode to cathode surface area was effective at the pilot-scale.
- The current density on the anode plates was found to be directly related to bromide ion removal at the pilot-scale. In general, higher current densities on the anode surface correspond to higher bromide removal. However, under lower HRT conditions (i.e., less than 3 minutes), due to the accumulation of bromine especially at higher current settings, the overall bromide removal was limited, and a threshold was observed, even when higher currents were applied.
- The power required to achieve a given percentage of bromide oxidation to bromine increased with decreasing HRT.
- Under high flow conditions (i.e., high turbulence), even though a lower HRT would decrease the electrolytic reactor size and required footprint, it would also require higher power. In addition, the overall benefits in terms of bromide removal and DBP reduction are more limited.
- The limited data obtained from sampling over the depth of the DSA plates showed that most of the bromide oxidation was occurring in the first 10 cm of the 3 ft depth. This finding would result in significantly lower DSA surface area requirements (approximately 1/10). The active HRT over the DSA depth would also be significantly decreased. However, a higher power would be required.
- Previous studies conducted by Kimbrough and Suffet (2002 (a) and (b)) showed that air stripping as a second stage was not effective for the removal of bromine. The majority of the bromine removal was achieved within the reactor.

A direct scale-up from pilot-scale to full-scale was performed based on the ratio of required anode surface area to raw water flow (m^2/lpm), and the required current density on the anode surface (amps/cm^2).

Assuming the entire DSA depth (i.e., 3 ft) is used, the DSA surface area used in the pilot plant would be 11,232-sq-in. The optimal conservative combination of flowrate and current were used to calculate the ranges of required DSA surface area to flow and required current density to DSA surface area (Table 6.1). The ratio of anode surface area to raw water flow ranged from 0.05 to 0.38 m^2/lpm during pilot testing, the current density on the anode surface ranged from 0.0005 to 0.0032 amps/cm^2.

The limited depth sampling results showed that a 35-amp current and 1/10 of the DSA depth were required under a 2-pm flowrate condition to obtain bromide removals equivalent to those under the same flowrate, using, 15 amps of current, and the entire 3 ft of DSA depth. Results are included in Table 6.1.

It should be kept in mind that bromide removal rates under higher turbulence conditions (i.e., 4 to 15 gpm) were not optimal. Therefore, for design purposes, the focus is placed on the data collected under a 2-pm flowrate, assuming both conditions of the entire and 1/10 of DSA depths.

The required anode surface area for a full-scale system was estimated by multiplying the optimum anode surface area to flow ratio from pilot testing by the full-scale flow. The current requirements were estimated by multiplying the optimum current density from pilot testing by the estimated full-scale anode surface area requirement. Once again, it is noted that the ratios from Table 6.1 were used to scale-up power and space requirements. However, the power and space requirements calculated under the 2-gpm flowrate should be used for design purposes. Both the full depth and 1/10 of DSA depth conditions were used.

In order to calculate the full-scale power requirements (Table 6.2), it was necessary to assume a voltage for a full-scale system. Typically, in a full-scale operation, the voltage would be held constant and the current allowed to vary. During pilot testing, because of the limitations of the equipment, the applied voltage varied along with the current, from approximately 15V to over 80V, with optimum conditions occurring between 15 and 25V for the low power testing. For the purpose of estimating full-scale power consumption, an intermediate value of 20V was assumed. The voltage can be set just above the anodic potential to oxidize bromide to bromine and the current allowed to vary. Further research into full-scale system performance, power consumption, and equipment needs at various voltages is recommended.

Table 6.1
Required design ratios extracted from pilot-scale data

Pilot-scale flow (gpm)	HRT (min)	Pilot-scale current (A)	DSA surface area to flow (m^2/lpm)	Current density to DSA surface area (amps/cm^2)
2*	11.2	15	**0.38**	**0.0005**
2†	1.2	35	**0.04**	**0.0112**
4	5.6	25	0.19	0.0009
5	4.5	61	0.15	0.0021
10	2.2	61	0.08	0.0021
15	1.5	90	0.05	0.0032

Note: Bolded values represent optimal conditions for bromide removal and bromine accumulation. Bolded ratios should be used for design purposes assuming full DSA depth and 1/10 DSA depth

* Entire 3 ft DSA depth used

† The first 10 cm of DSA depth used

Table 6.2
Scale-up of electrolytic cell

Pilot-scale flow (gpm)	HRT (min)	Pilot-scale current (A)	Full-scale anode surface area (m^2)	Full-scale current (A)	Full-scale power (kw)*	Full-scale basin volume (m^3)
2	11.2	15	40,000	208,000	4,200†	1,200†
2	1.2	35	4,300	486,000	9,700‡	126‡
4	5.6	25	20,000	174,000	3,500	589
5	4.5	61	16,000	339,000	6,800	471
10	2.2	61	8,000	169,000	3,400	236
15	1.5	90	5,200	167,000	3,300	157

* Scale up based on a 105,000 Lpm (40 mgd) system operating at 20 V potential.
† Design values assuming full depth of DSA plate
‡ Design values assuming 1/10 depth of DSA plate

The estimated full-scale basin volume was estimated by multiplying the HRT by full-scale flow, which was assumed to be 105,000 Lpm (40 mgd). Table 6.2 summarizes the optimal conditions for bromide removal at a range of flow rates, and the corresponding full-scale current, power, and basin volume requirements.

Geometry and Space Requirements of Full-Scale System

Early on, the space requirements for a full-scale system were recognized as a possible limitation on scalability. The pilot-scale reactor consisted of a rectangular basin that contained solid 12-in x 36-in DSA plates oriented parallel to one another, spaced four inches apart on center. Preliminary investigations found that a scale-up of the pilot reactor using the same spacing and entire DSA plate depth would be extremely large, resulting in an HRT of over 30 minutes. In order to achieve a lower HRT (smaller footprint), two conceptual designs were developed that allow for more anode surface area to fit within a given volume:

- A shallow open channel design that contains modules of DSA plates, wires, or ribbons (Figure 6.1).
- A deep basin design that contains submerged modules of DSA wires or ribbons (Figure 6.2).

The manufacturer of the DSA plates confirmed that DSA anodes are available in a number of configurations that may be compatible with the conceptual designs, including plates of various sizes, 1/4" and 1/2" diameter wires, and 1/4" and 1/2" wide ribbons. Furthermore, expanded mesh ribbons are available that provide an additional 15 percent more surface area per unit volume than solid ribbons.

The theoretical HRT of each of the two conceptual designs was calculated based on a spacing of plates/ribbons/wires of 0.5 inches. Square DSA plates measuring 36 inches by 36 inches were assumed for the scale-up. The HRTs of the conceptual designs were calculated by dividing the minimum volume needed to fit the required number of plates by the full-scale flow of 105,000 Lpm (40 mgd). Based on this plate size and spacing, and assuming the entire depth of DSA plate used at the pilot scale, HRTs as low as 3 to 6 minutes may be possible, therefore, reducing the footprint of a full-scale system. Assuming that the actual required DSA depth was 1/10 of entire depth at the pilot scale, HRTs of 1 min and lower may be possible. The placement of anodes in a full-scale system will be affected by the efficiency of the anode material in different configurations. The shape and surface area of the electrodes and the distance between them could have a significant impact on current density observed at different points along the surface. Wires and ribbons would have a lower voltage drop per unit length than plates and, therefore, may provide more flexibility in system geometry. Further research needs to be done to evaluate the performance of the electrolytic cell with various anode configurations in order to develop a full-scale design.

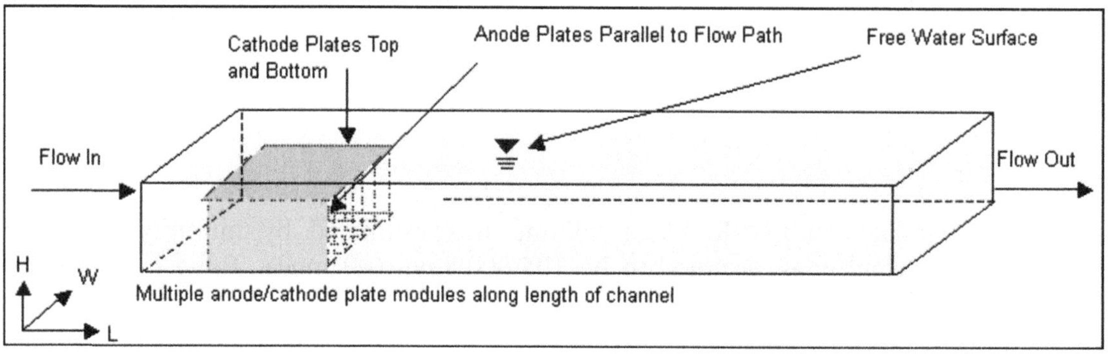

Figure 6.1 Shallow channel design for electrochemical cell reactor

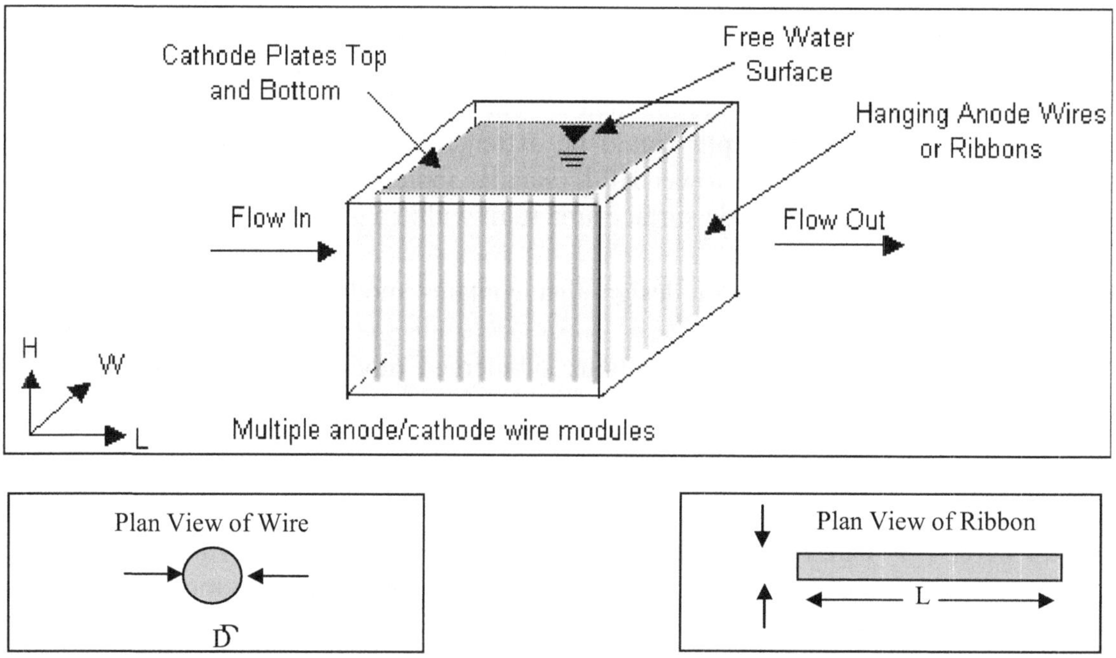

Figure 6.2 Submerged ribbon or wire module design for electrochemical reactor

Even under HRT values of 3 to 6 minutes, a full-scale shallow channel system would occupy a large footprint. For example, a 3-meter wide by 1-meter deep channel with an HRT of 3 minutes would be 11 meters in length. At an HRT of 6 minutes, the same channel would be 22 meters in length. The deeper submerged channel design would provide the smallest footprint, if anodes could be found to accommodate it. However, if the full-scale design is based on the 1/10 DSA depth, then the channel length can be decreased at a minimum by an additional 20 percent, and the number of modules decreased at minimum from 21 to 17. These numbers are based on a 2-gpm flowrate. Higher flowrates would further decrease the channel size.

A key factor influencing the layout of a full-scale system is the relationship between bromide removal efficiency and HRT and mixing. At lower HRT values, water moves through the reactor more quickly, so there is less time for bromide oxidation and volatilization to occur. Pilot-scale data suggested that higher power is required to achieve measurable bromide removal as the HRT decreases. However, limited data also suggested that if mixing within the reactor is kept at a minimum, then lower HRTs are more feasible and do not require as high of power. Therefore, the design of a full-scale system would need to balance the system footprint and power requirements.

Power Requirements

Perhaps the biggest challenge for scalability of the electrolytic cell reactor is the power requirement at full-scale. The electrolytic cell operates at low voltages, but requires high current for effective bromide oxidation. As shown in Table 6.2, the direct scale up of the pilot system would require power supplies of approximately 3 to 7 Megawatts at full-scale. As a basis of comparison, an ozone system for a similar sized facility (i.e., 40 mgd) would require on the order of 0.2 Megawatt of power for a 3.0-mg/L ozone dose.

Ozone systems operate at comparatively higher voltages and lower currents than an electrolytic cell. Finding a power supply that can output the high current at low voltage required by a full-scale electrolytic cell is likely to be extremely difficult. It will almost certainly be necessary to use multiple power supplies in parallel in order to operate such a system. For comparison purposes, the transformer/rectifier used in the OSEC B4-200 hypochlorite generation system, where it performs the similar function of providing a DC power supply to an electrolysis process, can supply 3,400A at 32V, and is approximately 1 meter wide, 1 meter deep, 1.5 meters tall, and weighs 885 kilograms. As many as forty such units could be required to operate a full-scale system. Research would have to be done to determine if there are power supplies available that are better suited to the task. There are some fundamental power supply limitations such as heat dissipation and conductor current density; however, that limits power supply options.

Another hurdle posed by the power requirements of a full-scale system are related to power distribution. The power supplies would have to be served at a medium voltage (4,160V or 12,470V) and extra transformers would have to be installed to step down the medium voltage supply to the lower voltage used by the electrolytic cell. Second, power must be transmitted from the power supplies to the electrodes themselves. Standard industrial cables are not capable of carrying the currents required for the full-scale system, so copper bus bars would be needed to deliver the currents required. The bus bars would not simply attach to the anode and cathode plates – this level of current capacity would have to be maintained by the entire circuit system, so if anodes and cathodes are bonded together rather than individually connected back to the power supply, the bonds between them would have to be rated for high current as well. All in all, the power delivery equipment would need to be extremely specialized for this application and would likely pose a significant cost.

Maintenance

The potential maintenance requirements for a full-scale system would likely include the regular cleaning of the anode surface. The DSA manufacturer (ELTECH) has indicated that anodes used for this type of application may benefit from regular cleaning consisting of a pressure wash (to remove dirt/debris, accumulation of biological material), and intermittent cleaning to remove scaling due to cations in the raw water (calcium, magnesium, lead, etc.). Intermittent cleaning of the anodes may be performed using chemicals or reverse polarity. Maintenance of anode modules may be designed to operate similar to the maintenance of submerged membrane modules, whereby modules are periodically removed for cleaning before being returned to service. Further research is needed to estimate the required cleaning methods and intervals for a full-scale system.

Safety

Care would need to be taken to isolate all electrical components and prevent human contact. The power supply equipment would require high voltages so appropriate safety precautions would be required around that equipment. The voltage in the system would likely be relatively low and not pose a shock threat, but components carrying high currents can run at extremely high temperatures, creating the potential for serious burns. As the operating voltage of the system is determined, consideration should be given to the potential for injury and how to prevent human contact with the system while in operation.

In addition to practical safety, it must be ensured that the system meets applicable electrical and safety codes, such as the National Electrical Code, National Electric Safety Code, OSHA, etc. This may present a challenge, as applying very high currents to open channels of water presents significant safety concerns. The high currents required by a full-scale system could travel long distances through the water, so it would be necessary to evaluate the effects of the high current on upstream and downstream processes. Further research would need to be done to evaluate how far the current may travel and its potential effects.

CONCEPTUAL DESIGN PARAMETERS

The conceptual design parameters associated with channel, submerged wire, and submerged ribbon concepts are summarized in Table 6.3.

CONCEPTUAL CAPITAL COST ANALYSIS OF OPEN CHANNEL DESIGN

A conceptual capital cost analysis was conducted assuming a 40-mgd open channel design. Only key items are included. Capital costs are summarized in Table 6.4. Both the full depth and the fraction of DSA depth were considered in the analysis. Electrical cost is a significant component of the total cost. The cost was based on 40 power supply modules at $33,000 per module, $127,000 for a 480 V, 5000 A Switchboard with 40 feeder breakers, $120,000 for a transformer, $600,000 for wires, conduits, Bus bars, upgrades, and another $650,000 of labor costs.

Table 6.3
Conceptual design parameters

Parameter	Channel design	Submerged wire design	Submerged ribbon design
Flowrate (mgd)	40	40	40
Anode surface area to flow (m^2/lpm)	0.38* 0.04†	0.43* 0.05†	0.43* 0.05†
Current density to anode surface area (Amps/cm^2)	0.0005* 0.0112†	0.0005* 0.0097†	0.0005* 0.0097†
Design anode dimensions	3 ft x 3 ft	5 ft high, 0.125-in diameter	5 ft high, 0.04-in wide, 0.5-in length
Number of anodes/module	150	25,600	17,818
Number of required modules	158* 17†	117* 13†	58* 6†
Channel width (ft)	11	8	8
Channel length (ft)	553* 61†	881* 96†	432* 47†
Channel depth (ft)	5	7	7
HRT (min)	4.9* 0.5†	9.5* 1.0†	4.7* 0.5†
Total required anode area (m^2)	40,000* 4,400†	46,000* 5,000†	46,000* 5,000†
Total required cathode area based on channel design (m^2)	1,130* 125†	1.310* 143†	642* 70†
Design required current (Amps)	208,000* 486,000†	208,000* 486,000†	208,000* 486,000†

*Based on full DSA depth, 2 gpm, 15 amps
† Based on 1/10 DSA depth, 2 gpm, 35 amps

Table 6.4
Conceptual capital costs for 40-mgd electrolytic open channel reactor

Cost variable	Capital cost based on full DSA depth	Capital cost based on 1/10 DSA depth	Assumptions
Open channel reactor	$600,000	$100,000	Concrete cost only No internal baffles; open top
Cost of DSA plates per module	$753	$753	Based on $3.0/sq-ft area (high-end cost)
Total DSA plates cost	$1,280,000	$140,000	
Total SS Cathode Plates Cost	$164,240	$18,200	Based on $13.5/ sq-ft area
Electrical equipment	$2,800,000	$2,800,000	
Instrumentation & control	$205,000	$26,000	10% of total cost
Subtotal	$5,050,000	$3,085,000	
Engineering & admin	$1,010,000	$617,000	20% of subtotal
Contingency	$1,515,000	$925,000	30% of subtotal
Total cost	$7,575,000	$4,627,000	
Amortized total cost (assuming 20 years, 5%)	$608,000	$371,000	

O&M COST FACTORS

DSA Plates O&M Cost Factors

O&M costs related to DSA plates would likely include cleaning/flushing on an intermittent basis to remove miscellaneous deposits and acid wash to remove hardness scale buildup. Other costs include the replacement of plates every 5 to 10 years. Insufficient information is available at this time to estimate these costs.

Power Costs

Preliminary power costs in $/MG treated water were calculated for each optimal electrolysis condition from pilot-scale data (Table 6.5). The power costs were calculated in two ways: 1) extrapolated directly from pilot-scale data, and 2) assuming a constant potential of 20 V to mimic full-scale operation. The optimal electrolysis condition was based on a combined maximum percent bromide removal and minimum accumulated bromine residual. In certain cases, optimal conditions were not achieved (requiring either higher or lower power than tested). The power cost was based on a $0.1/kw-h value.

Table 6.5
Ranges of power costs in $/MG associated with optimal electrolysis conditions

Flow (gpm)	HRT (min)	Pilot-scale extrapolated ranges of power cost ($/MG)	Extrapolated ranges of power cost assuming constant potential of 20 V ($/MG)
2*	11.2	170	250
2†	1.2	398	583
4	5.6	213	208
5	4.5	213<<895	407
7	3.2	>270 to <430	
10	2.2	>270<430	200
15	1.5	271 to 640	200

*Assuming full DSA depth, 15 Amps
† Assuming 1/10 DSA depth, 35 Amps

It appears that power costs increase with increasing flow rate/turbulence. The bromide oxidation is rate dependent, and the reaction is more power-intensive under high-rate and high turbulence/velocity conditions. When a fraction of the DSA depth is assumed, the power it takes to obtain efficient bromide removal is higher, resulting in higher power cost. As such, operational costs will need to be balanced with the capital costs and overall required footprint. Under full-scale conditions and constant potential, the power cost in $/MG was more or less constant over the range of flowrates/HRTs. Assuming $200/MG, running a 40-mgd electrolytic system would cost $8,000/day or $2,920,000 per year. This cost is high, and more research is required to assess whether the power costs can be decreased.

CONCLUSIONS

A conceptual scale-up was conducted involving three types of reactor configurations: open-channels incorporating DSA plates, DSA ribbons, and DSA wires. Depending on the HRT and current conditions, the ratio of required DSA surface area to flowrate ranged from 0.05 to 0.38 m^2/Lpm. The current density on the DSA plates ranged from 0.0005 to 0.0032 amps/cm^2. When a fraction of the DSA plate depth was assumed, the ratio of requires DSA surface area to flowrate was 0.04 m^2/Lpm, whereas the current density was more in the order of 0.0112 amps/cm^2. Corresponding full-scale (i.e., 40 mgd) power requirements and required basin volumes were calculated. Because the 2-gpm flowrate was observed as optimal in terms of removing bromide, the current density and required anode surface area/flow associated with that flowrate, under both full and partial DSA depth assumptions, were used for design purposes. Optimizing the configuration of the DSA plates or ribbons/wires resulted in decreased HRT/footprint requirements. However, even at those lower HRTs, a large footprint would be required. As discussed in Chapter 4, limited data has shown that the majority of beneficial bromide oxidation was occurring over the first 10 cm of the DSA plates depth. Above that, bromine is being accumulated. This could lead to significantly smaller HRT, footprint and volume, and DSA surface area requirements. Further testing should investigate ways to reduce the current requirements in a full-scale system and optimize the reactor design and anode configuration.

A direct scale-up to 40 mgd (105,000 Lpm) would require power supplies of 3 to 7 Megawatts. Finding a power supply that can output a high current at low voltage is likely to be extremely difficult and would require multiple power sources (as much as 40 units). Research would need to be conducted to determine the availability of power supplies better suited this task. Power distribution will also require specialized equipment and will likely pose a significant cost.

Safety and maintenance issues will also need to be further investigated. Applying high currents to open channels of water presents significant safety concerns. Effects on downstream processes will need to be identified.

System costs in the order of $4,627,000 to $7,575,000 were calculated. The lower cost was calculated assuming a fraction of the DSA depth. Electrical equipment forms a large proportion of the capital cost. O&M costs will likely include power costs, maintenance costs, and safety related costs. Power costs were estimated at $200/MG, resulting in $8,000/day for a 40-mgd system. Total annual costs for installing a 40 mgd electrolytic reactor are estimated at $3.5 M (assuming the full length of DSA plates).

In comparison, the capital cost for installing reverse osmosis treatment for bromide removal is estimated at $22M. The cost for pH reduction for bromate control would run at approximately $500,000 per year. The cost for the addition of MIEX® process for DBP removal to a 40 mgd conventional water treatment train is approximately $1.8 M per year.

It is recommended to conduct further research to better define scaling factors, operational, maintenance, and safety constraints associated with the system. Several configurations of DSA anodes should be considered to minimize footprint and volume requirements.

CHAPTER 7
FUTURE NEEDS AND RECOMMENDATIONS

A round-table workshop was held towards the end of the project to present the findings, discuss data gaps, and identify future research needs. Topics discussed included:

- Defining the hydraulics within the reactor
- Identifying the effective field area over the length of the DSA plate
- Testing a narrower range of currents under various flowrate conditions
- Formation and qualification/quantification of gases formed
- Repeating heterotrophic bacteria regrowth studies under different quenching conditions
- Conducting additional bromate formation potential testing
- Identifying the geometry/size of gas bubbles formed
- Assessing the effect of ammonia

The recommendations of the panel along with the authors' recommendations are summarized in the next paragraphs.

REACTOR CONFIGURATION

The current design of the reactor incorporates 1 ft wide x 3 ft deep DSA plates. In Chapter 6 (Scale-Up Issues), two other conceptual designs were introduced to maximize the DSA surface area and decrease required current densities and footprint. Instead of plates, thin wires or ribbons were investigated. This follows the principle of membrane module design, in which several cartridges with packed membrane fibers are utilized. While it will be difficult to achieve that level of surface area per volume, a revised design and anode configuration should be investigated.

This alternative design could also address another issue: it is not quite understood to which extent the DSA plates surface areas were utilized. Does the reaction occur over the entire surface of the plate or just at the boundaries? A thinner DSA wire or ribbon design would address that issue.

Limited data has shown that the majority of beneficial bromide oxidation was occurring over the first 10 cm of the DSA plates depth. Above that, bromine was being accumulated. This could lead to significantly smaller HRT, footprint and volume, and DSA surface area requirements. Further testing should investigate ways to reduce the current requirements in a full-scale system and optimize the reactor design and anode configuration. Depth data should be taken over the length of the reactor to obtain the bromide and bromine profile data over the length of the DSA plates. Alternate configurations could include the following changes:

- Rotating the DSA plates 90° such that the DSA plates are 1 ft deep and 3 ft wide
- Changing the spacing of the DSA plates
- Get the DSA plates closer to the cathode plates
- Using DSA ribbons or wires instead of plates

REACTOR OPERATION

A wide range of current amperages were applied during the course of the testing. Under hydraulic retention times of 3.2 to 11 minutes, 5 to 35 amps of current were applied. Under hydraulic retention times of 1.5 to 4.5 minutes, the currents ranged from 61 to 120 amps. These current ranges were often too wide to allow the determination of the optimal conditions of electrolysis. Often, the optimal current would lie somewhere between currents or outside the current range.

It would be recommended to repeat the electrolysis experiments using a narrower range of applied current, under various conditions of flowrate. The existing pilot reactor could be used for that purpose.

HALOGENATED DBPFP TESTS

In parallel to the reactor runs, it is recommended to subject the raw and electrolyzed samples to further DBPFP testing. The samples should be split in halves, with only one of the halves receiving a dechlorinating agent prior to addition of the chlorine. This would shed some more light on whether the accumulated bromine is playing a detrimental role on the formation potentials.

To confirm the findings of the TOBr and TOCl formation potential tests, it is recommended to subject the raw and electrolyzed samples to TOX speciation analysis.

BROMATE FORMATION POTENTIAL TESTS

During the course of testing, a limited number of raw and electrolyzed samples were ozonated and analyzed for bromate. Limited amount of data showed that the bromate formation potentials were likely to decrease in electrolyzed samples, however this decrease was within the analytical limits.. It is recommended to repeat those tests to confirm the findings and to further define the optimal current condition for bromate control.

MICROBIAL REGROWTH

Additional heterotrophic bacteria regrowth experiments are recommended using a variety of quenching agents (e.g., ascorbic acid, sodium sulfite, sodium thiosulfate), and sterilization conditions (autoclave, filter sterilization). It was also recommended to measure AOC in the raw and electrolyzed samples to estimate the biological regrowth potential of the electrolyzed water.

ELECTROLYSIS DOWNSTREAM OF CFS

During the course of testing, the electrolytic reactor was tested upstream of ozonation and CFS to mimic the present configuration of several water treatment plants, including CLWA. It would be however of benefit to test the reactor downstream of CFS. Conventional treatment will likely remove a portion of the DBP precursors (e.g., TOC). Upon electrolysis and the formation of chlorine and bromine, the treated water will be less likely to form DBPs.

SCALE UP, MAINTENANCE, AND SAFETY ISSUES

Further investigation and additional testing should be conducted to address the following issues:

- Identify ways to reduce the current requirements in a full-scale system and optimize the reactor design and anode configuration
- Identify power consumption, and equipment needs at various voltages
- Evaluate the performance of the electrolytic cell with various anode configurations in order to develop a full-scale design. The shape and surface area of the electrodes and the distance between them could have a significant impact on current density observed at different points along the surface
- Finding a power supply that can deliver the high current at low voltage required by a full-scale electrolytic cell will be a challenge. Determine if there are power supplies available that are better suited to the task.
- Identify power delivery equipment specialized for this application and identify the costs
- Define the maintenance needs associated with the operation of the DSAs: e.g., intermittent cleaning using chemicals or reverse polarity, periodic removal of modules for cleaning before being returned to service. Estimate the required cleaning methods and intervals for a full-scale system.
- Evaluate how far the current may travel and its potential effects on safety.

It is, therefore, recommended to conduct further research to better define operational, maintenance, and safety constraints associated with the system. Several configurations of DSAs should be considered to minimize footprint requirements. The reactor should be run over a long period of time so that maintenance and operational constraints can be identified and defined.

COST TO BENEFIT ANALYSIS

A cost to benefit analysis should be conducted to assess the feasibility of electrolysis for bromide removal in a full-scale application. Tangent costs include capital costs of the anodes, cathodes, contactor, and required specialized electrical and instrumentation equipments. Other cost aspects include safety risks, and required footprint. Potential benefits include decreased halogenated DBPFPs, decreased bromate formation potentials, biostable effluent, and chloride removal.

There is a balance between cost of treatment and associated benefits. What is the minimum HRT and smallest size electrolytic system that can be used to decrease harmful DBPs, while keeping the capital and power costs as low as possible?

APPENDIX A
TOTAL BROMIDE, BROMIDE, AND BROMINE CURVES AS A FUNCTION OF POWER (LOW POWER TESTING)

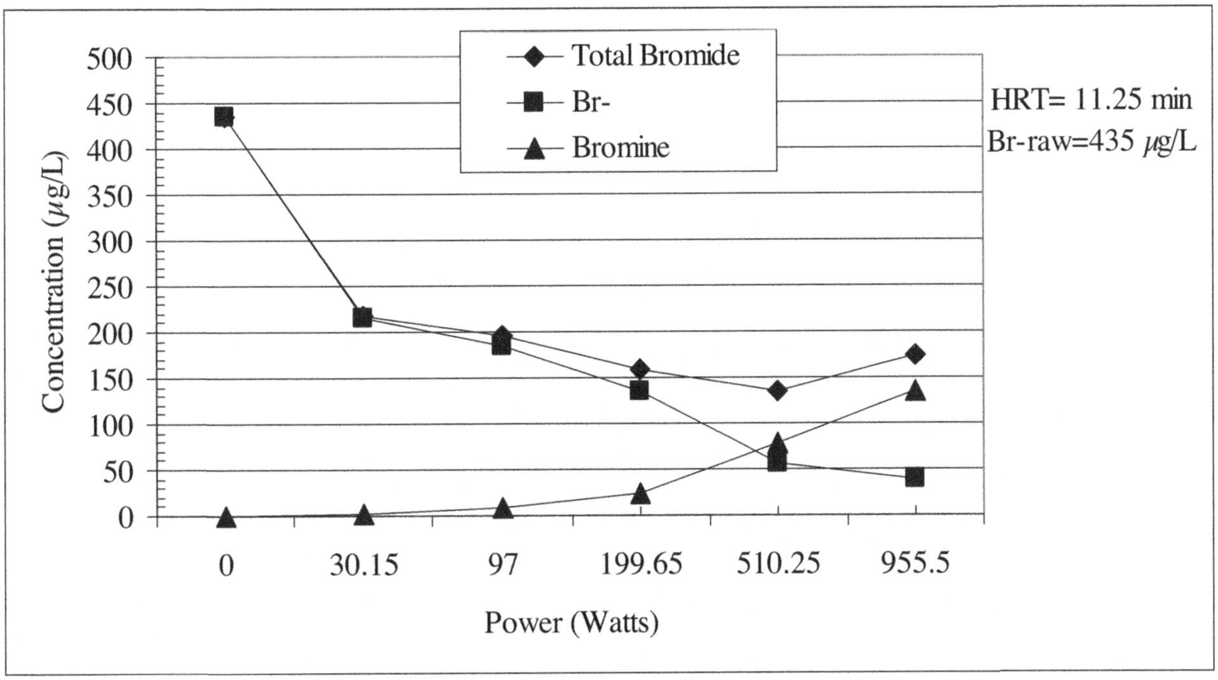

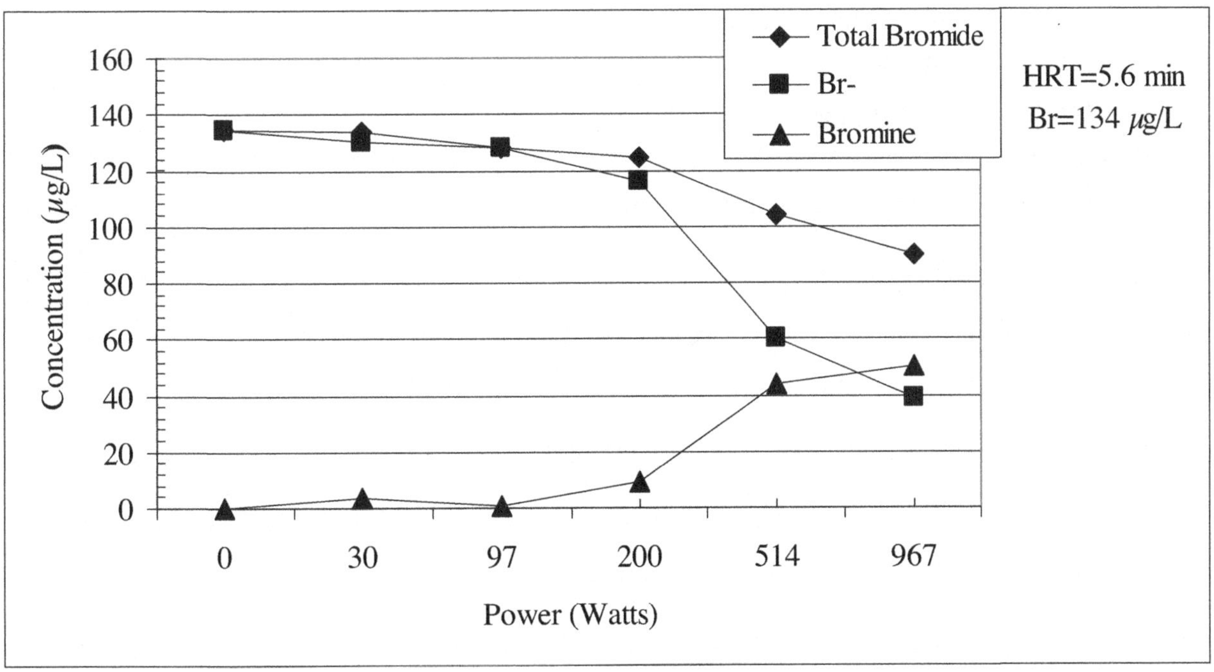

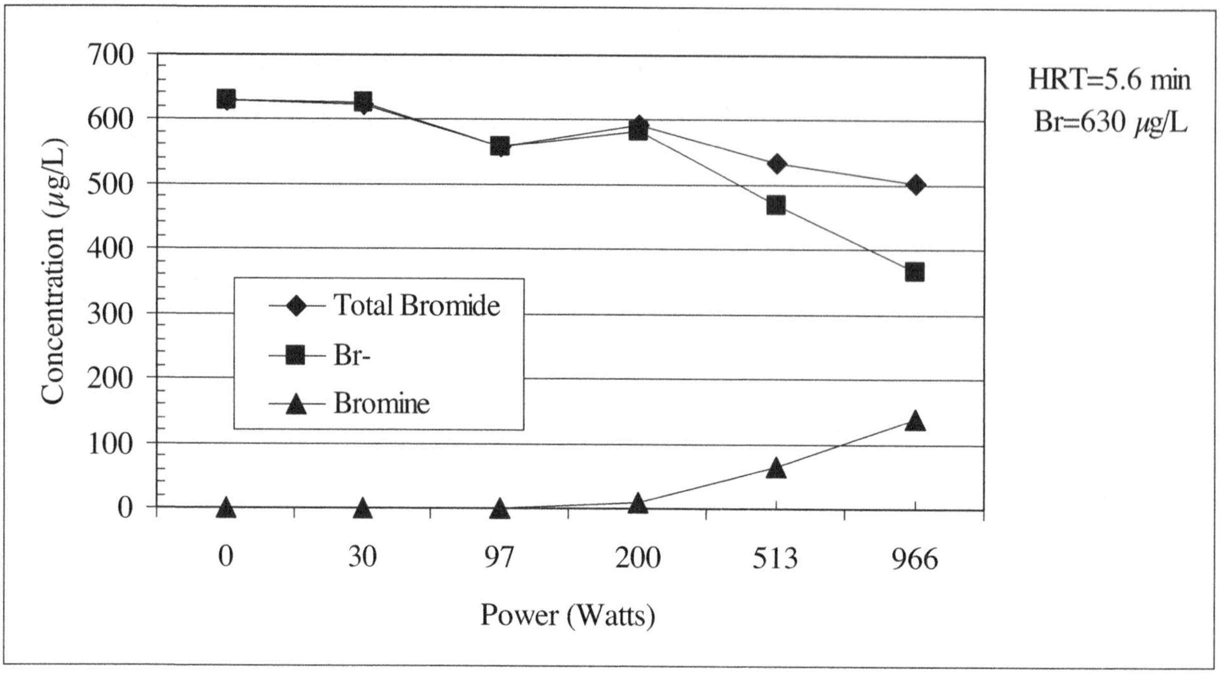

HRT=5.6 min
Br=630 μg/L
Concentration (μg/L)
700
600
500
400
300
200
100
0
Total Bromide
Br-
Bromine
0
30
97
200
513
966
Power (Watts)

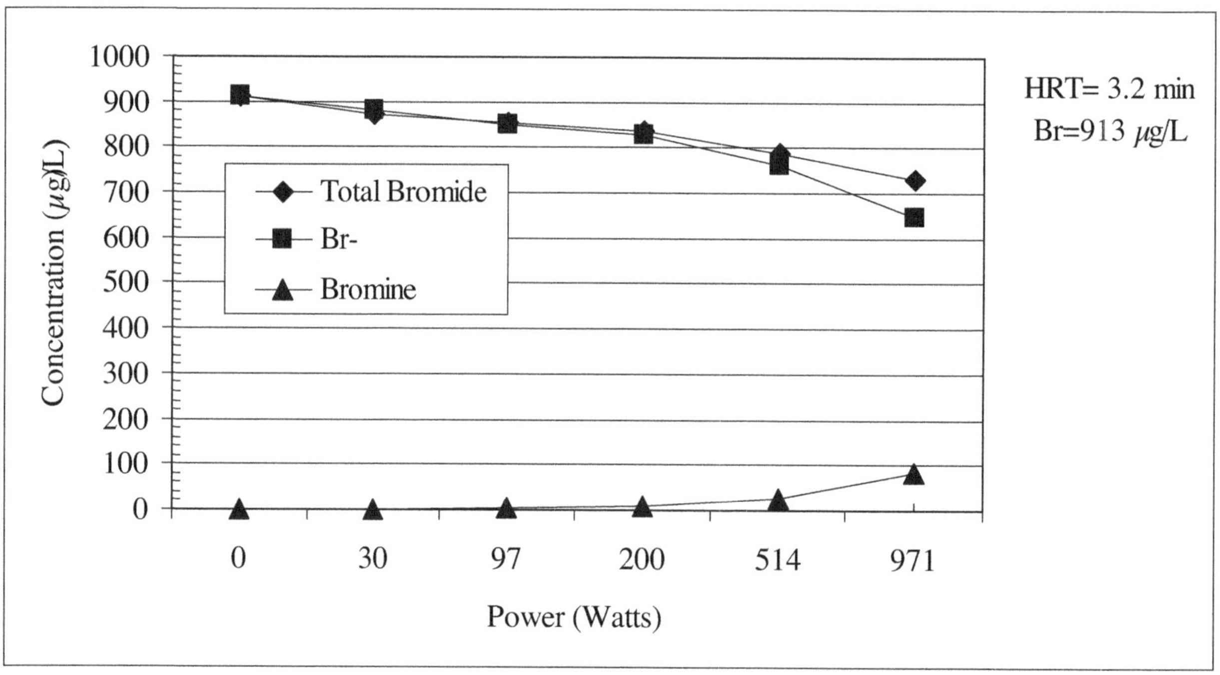

HRT= 3.2 min
Br=913 μg/L
Concentration (μg/L)
1000
900
800
700
600
500
400
300
200
100
0
Total Bromide
Br-
Bromine
0
30
97
200
514
971
Power (Watts)

APPENDIX B
TOTAL BROMIDE, BROMIDE, AND BROMINE CURVES AS A FUNCTION OF POWER (HIGH POWER TESTING)

15 GPM

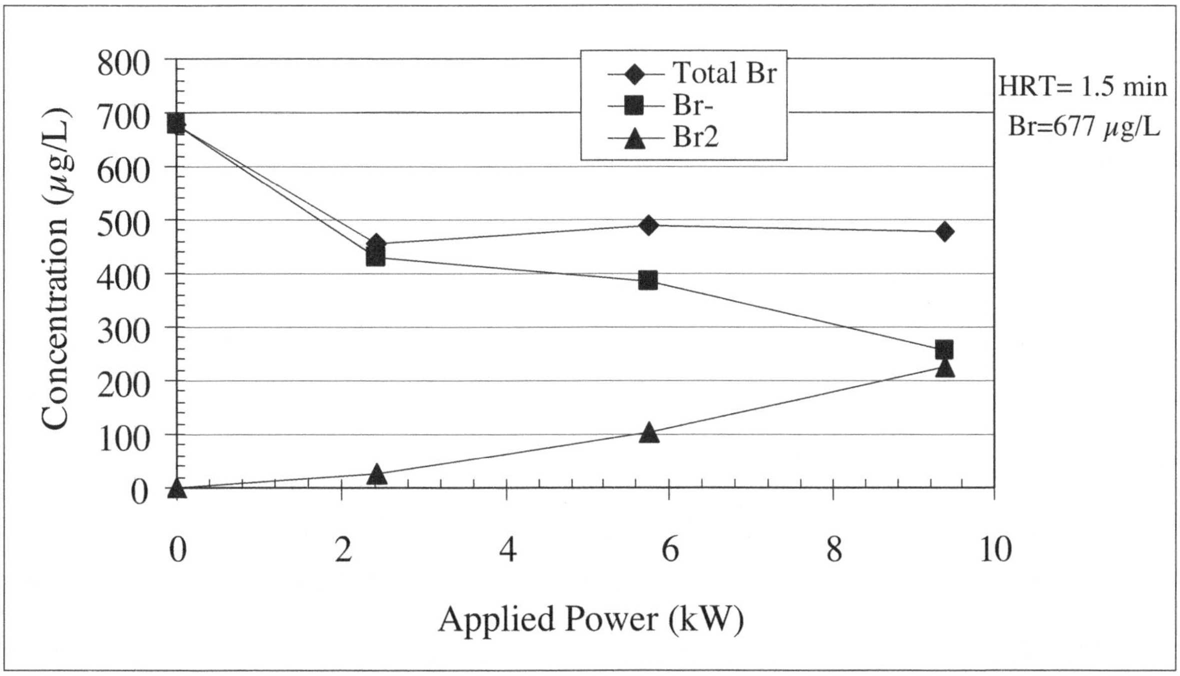

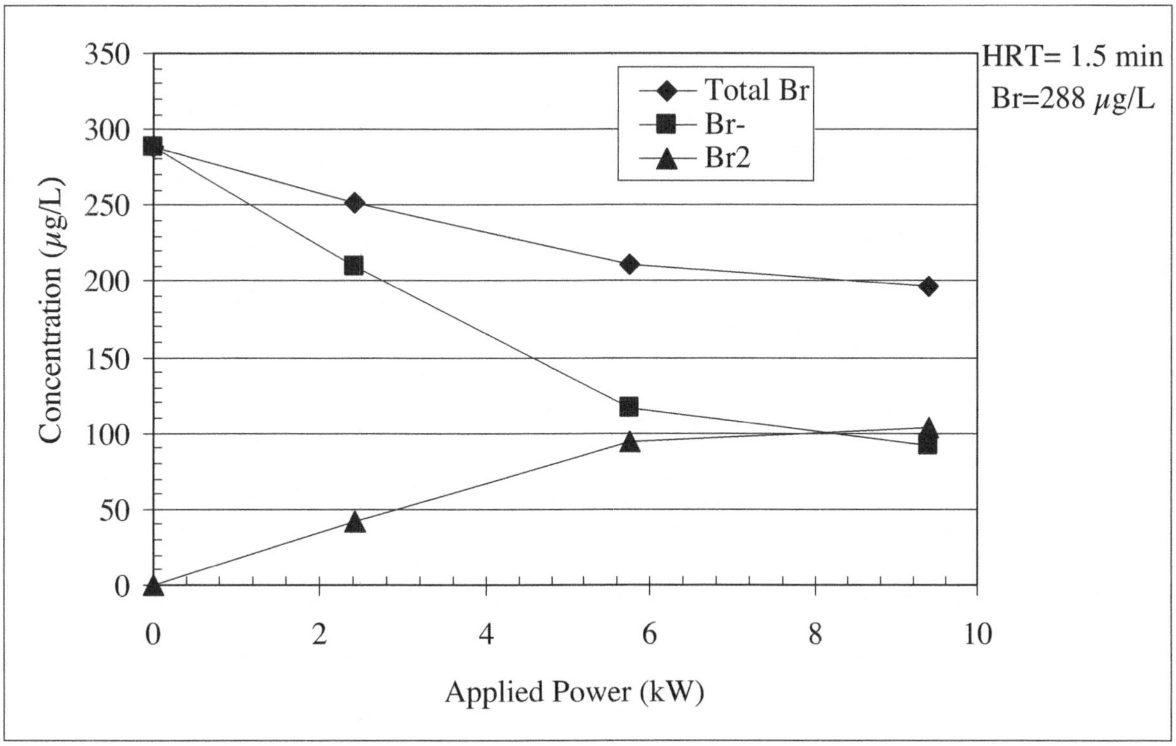

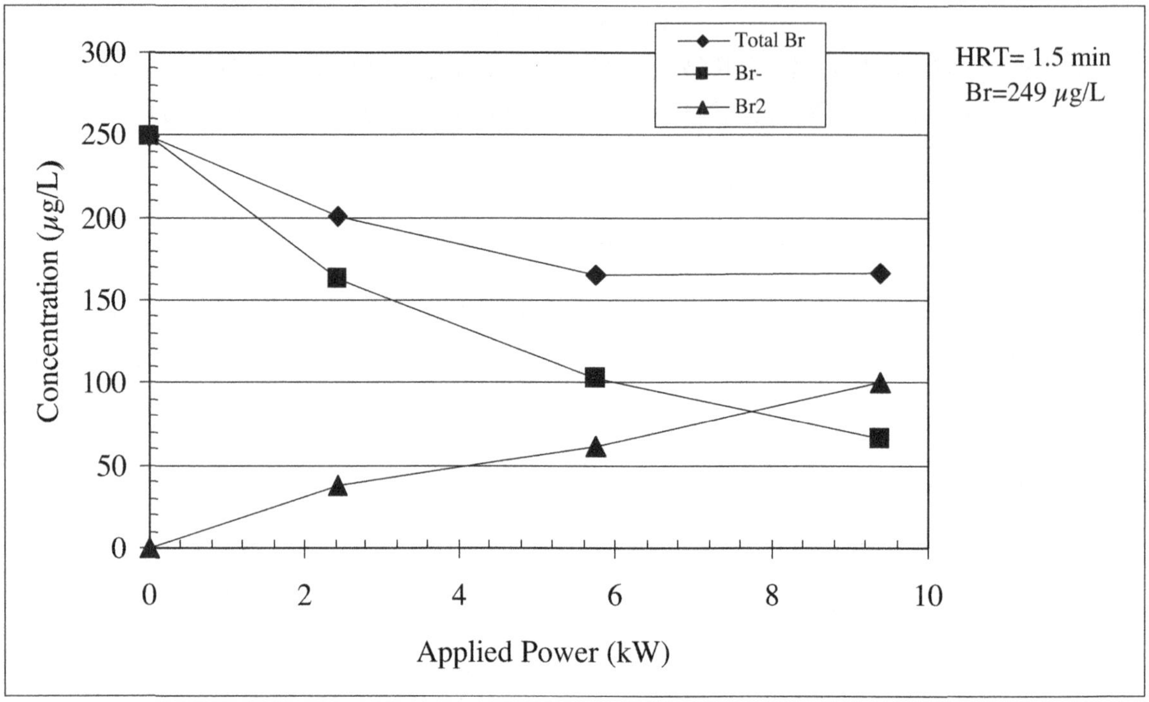

Total Br
Br-
Br2
HRT= 1.5 min
Br=249 μg/L
Concentration (μg/L)
Applied Power (kW)

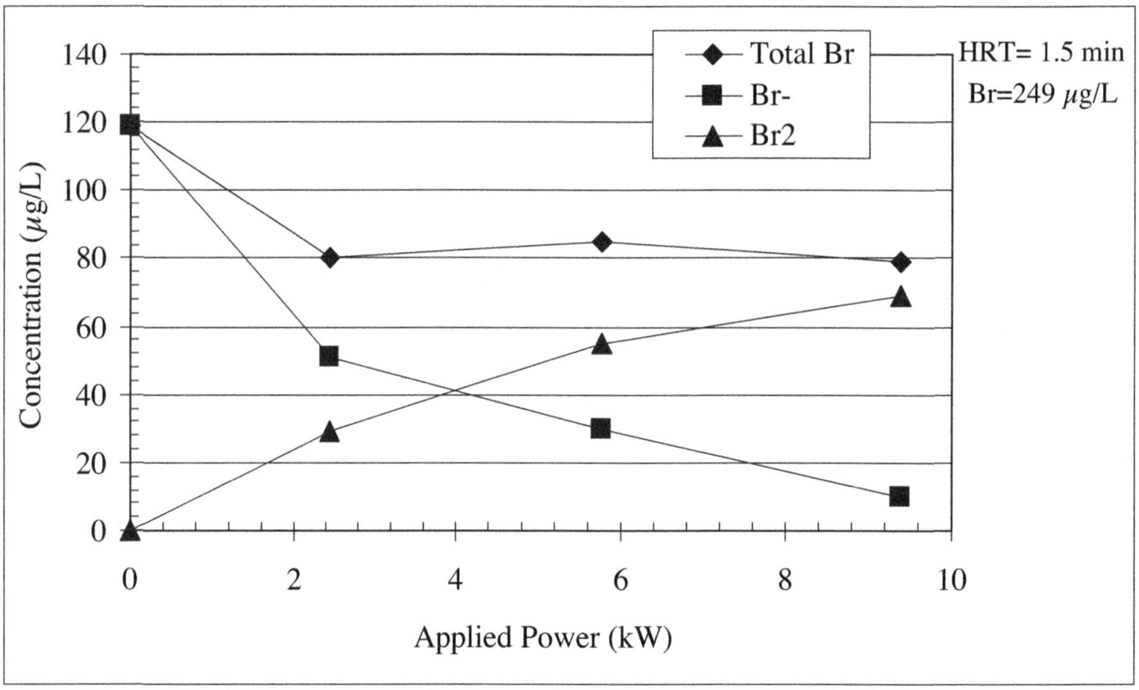

Total Br
Br-
Br2
HRT= 1.5 min
Br=249 μg/L
Concentration (μg/L)
Applied Power (kW)

10 GPM

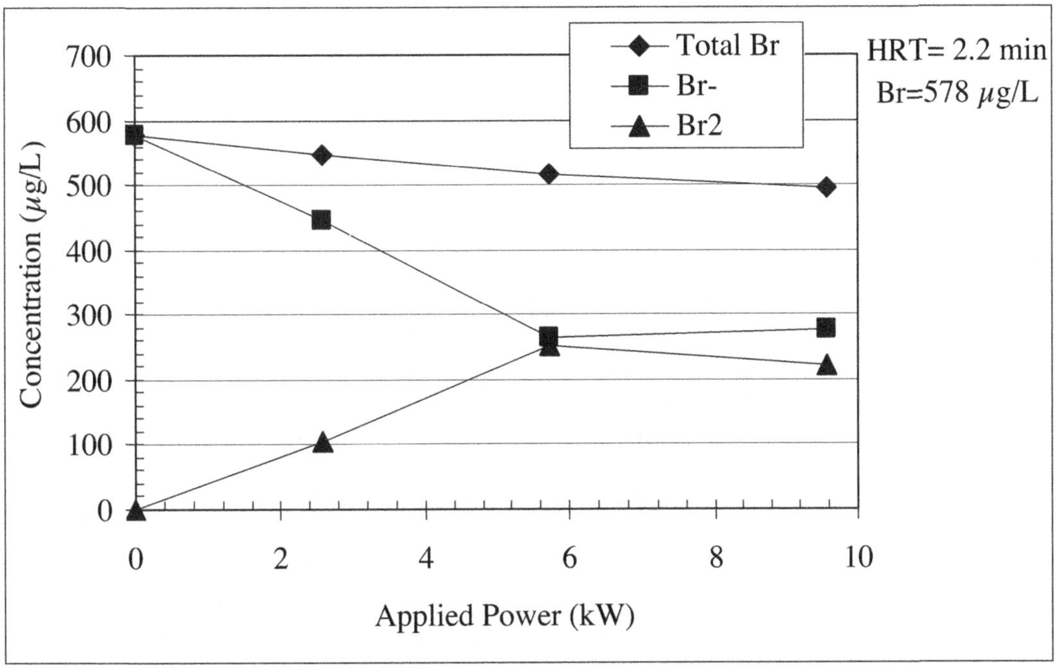

Total Br
Br-
Br2
HRT= 2.2 min
Br=578 μg/L
Concentration (μg/L)
700
600
500
400
300
200
100
0
0
2
4
6
8
10
Applied Power (kW)

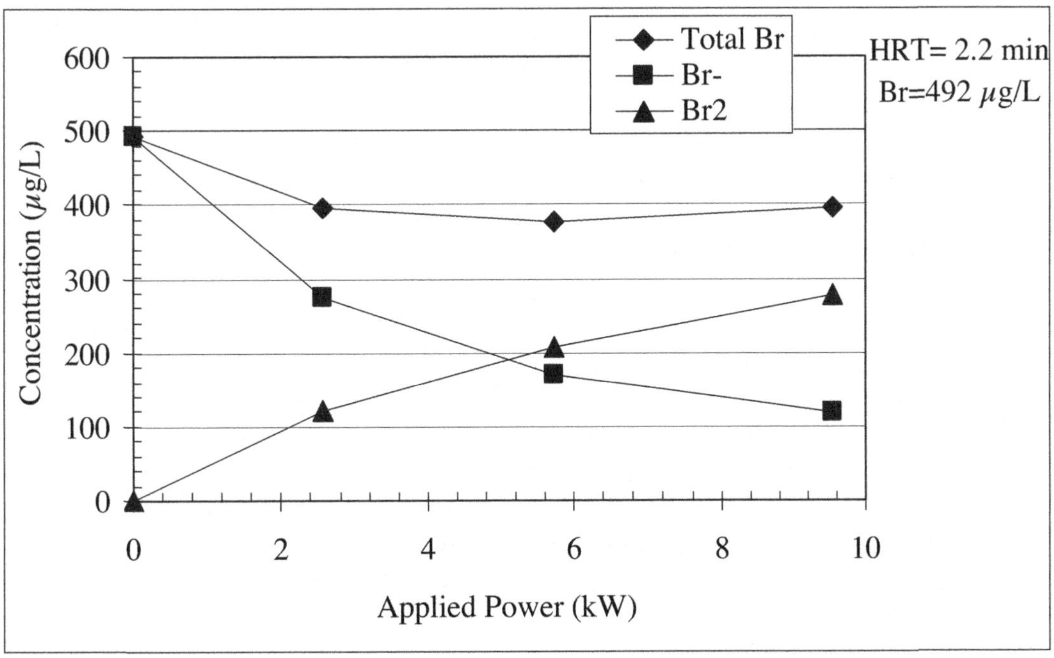

Total Br
Br-
Br2
HRT= 2.2 min
Br=492 μg/L
Concentration (μg/L)
600
500
400
300
200
100
0
0
2
4
6
8
10
Applied Power (kW)

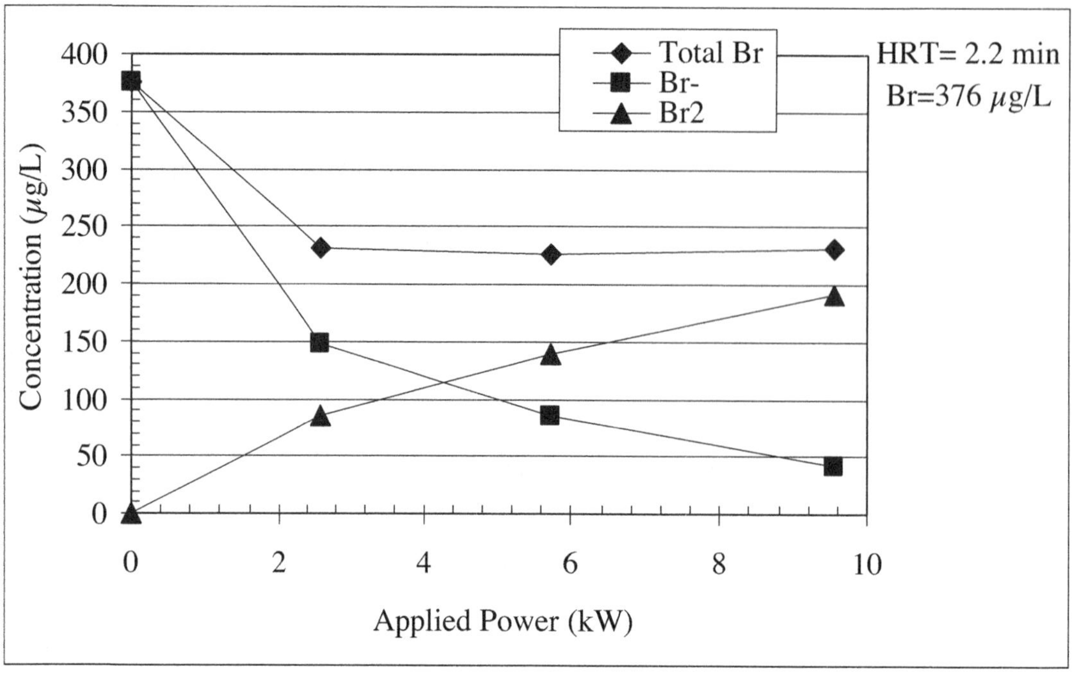

HRT= 2.2 min
Br=376 μg/L
Total Br
Br-
Br2
Concentration (μg/L)
400
350
300
250
200
150
100
50
0
0
2
4
6
8
10
Applied Power (kW)

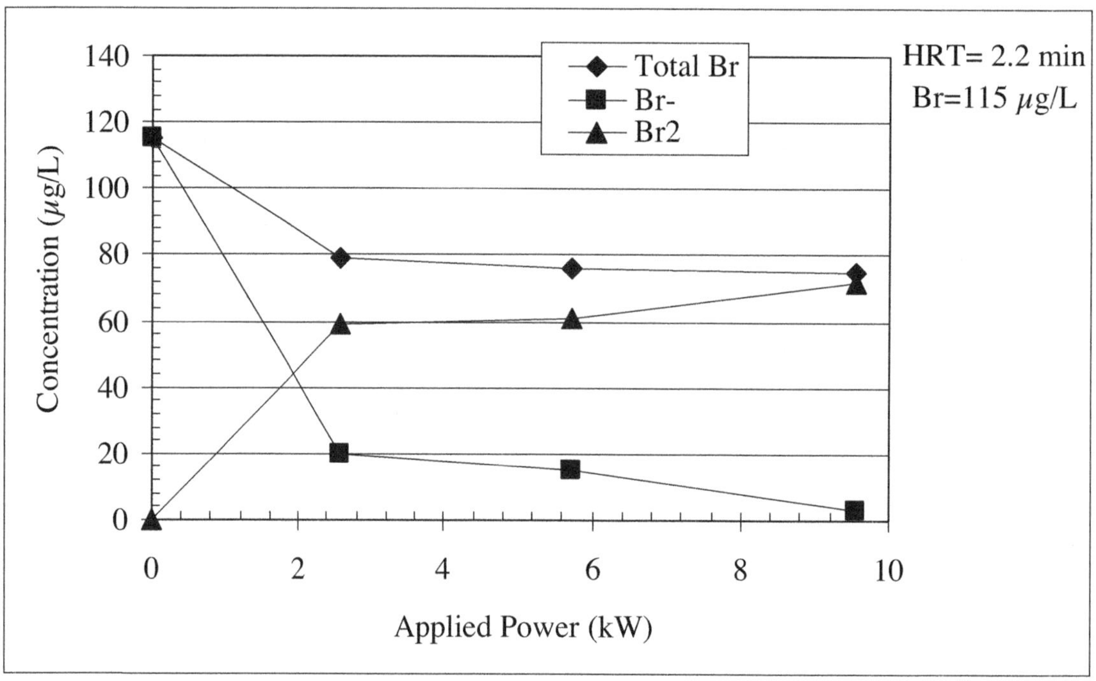

HRT= 2.2 min
Br=115 μg/L
Total Br
Br-
Br2
Concentration (μg/L)
140
120
100
80
60
40
20
0
0
2
4
6
8
10
Applied Power (kW)

5 GPM

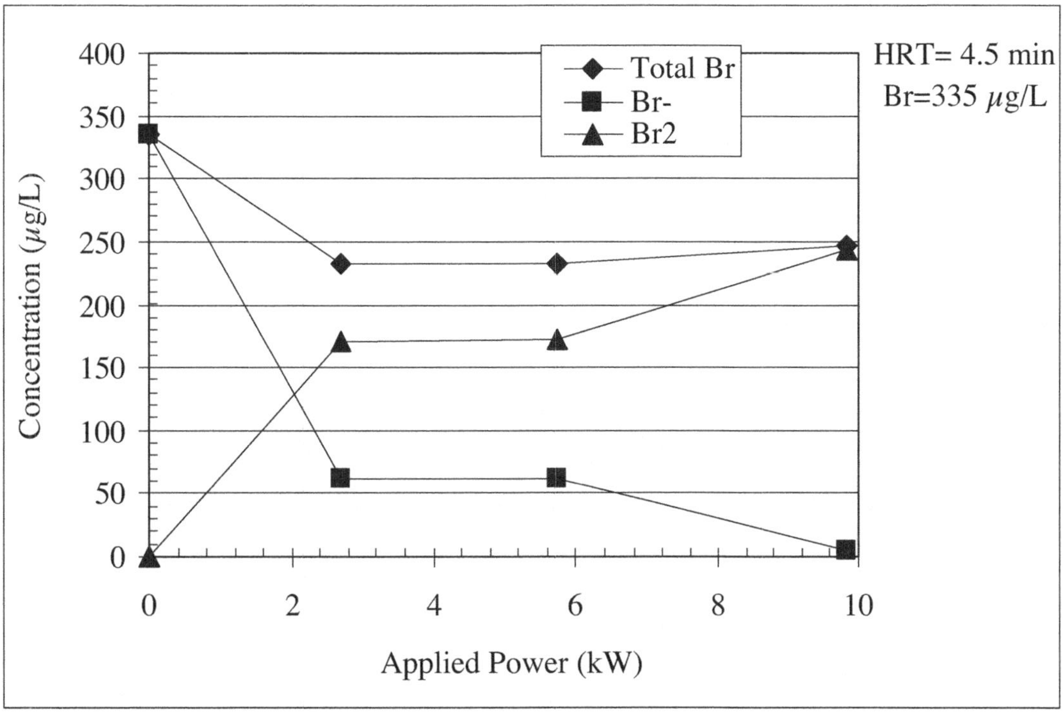

HRT= 4.5 min
Br=335 μg/L
Total Br
Br-
Br2
400
350
300
250
200
150
100
50
0
Concentration (μg/L)
0
2
4
6
8
10
Applied Power (kW)

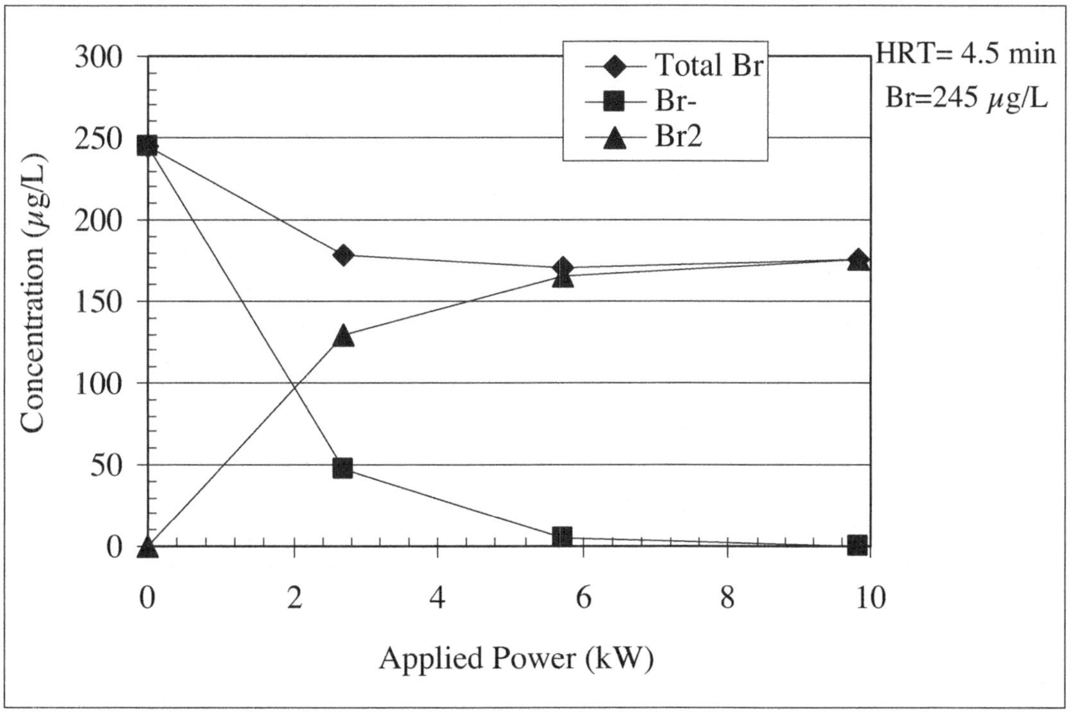

HRT= 4.5 min
Br=245 μg/L
Total Br
Br-
Br2
300
250
200
150
100
50
0
Concentration (μg/L)
0
2
4
6
8
10
Applied Power (kW)

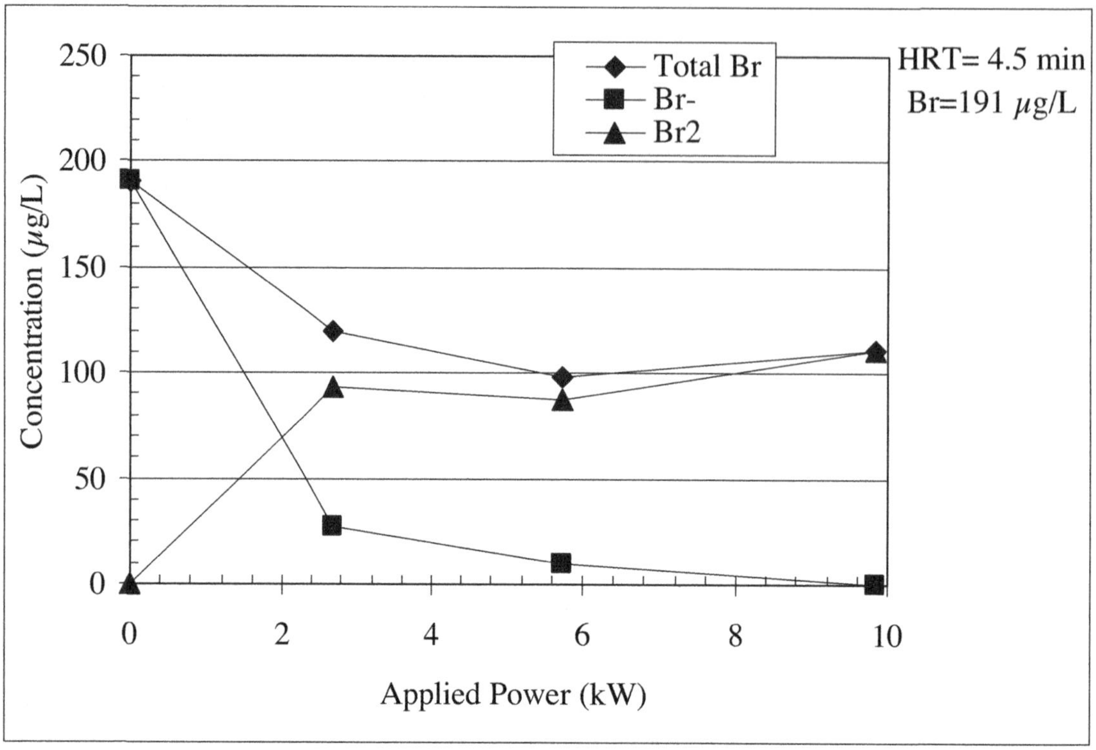

Total Br
Br-
Br2
HRT= 4.5 min
Br=191 µg/L
Concentration (µg/L)
250
200
150
100
50
0
0
2
4
6
8
10
Applied Power (kW)

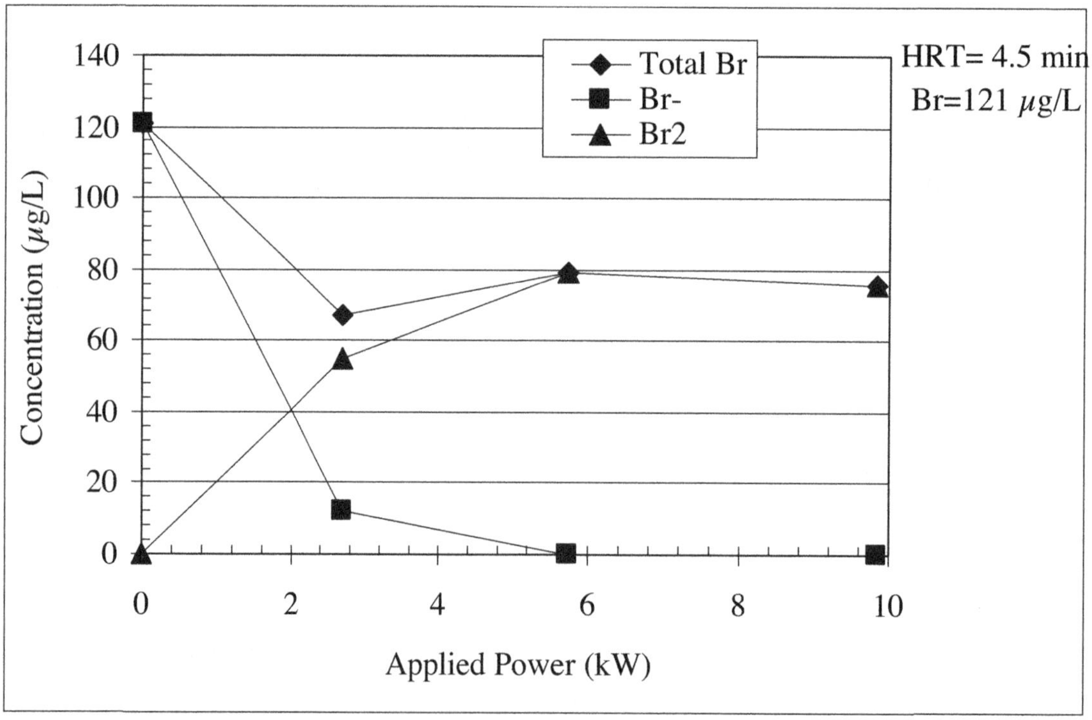

Total Br
Br-
Br2
HRT= 4.5 min
Br=121 µg/L
Concentration (µg/L)
140
120
100
80
60
40
20
0
0
2
4
6
8
10
Applied Power (kW)

APPENDIX C
BROMIDE REMOVAL AND BROMINE ACCUMULATION (HIGH POWER TESTING)

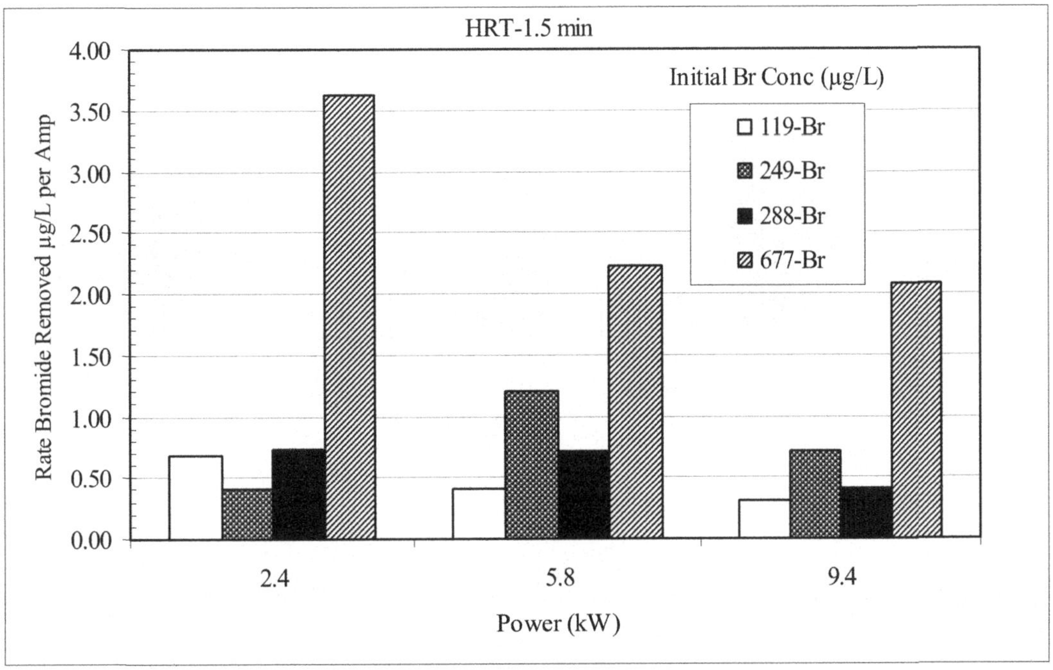

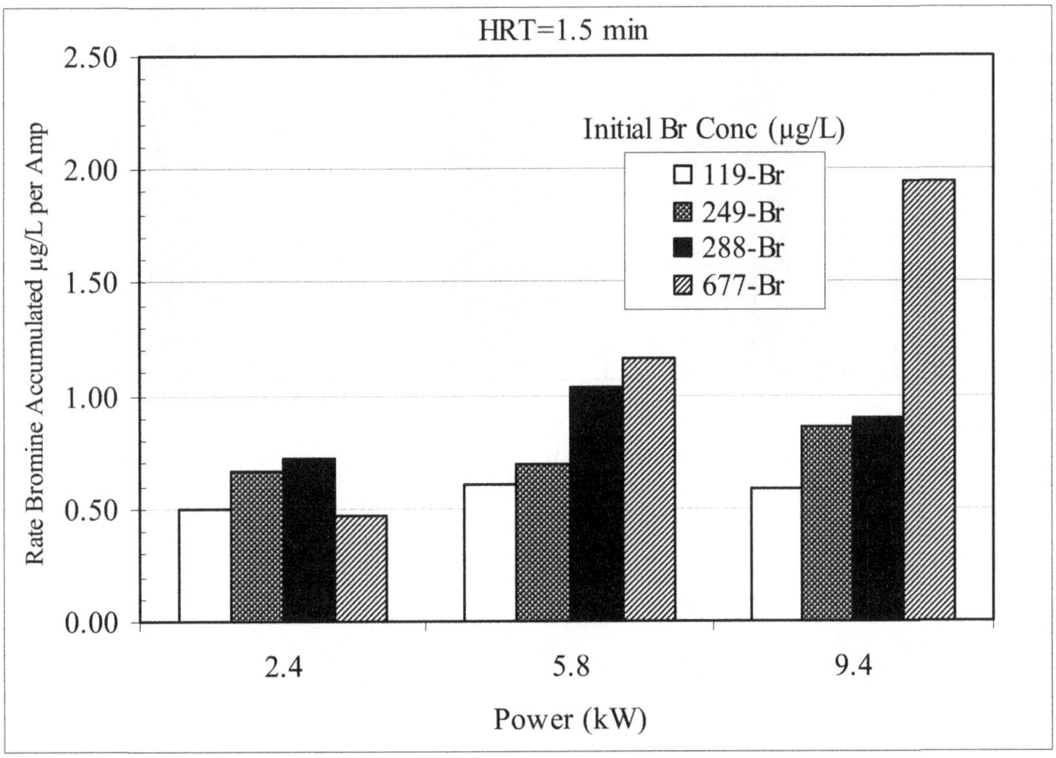

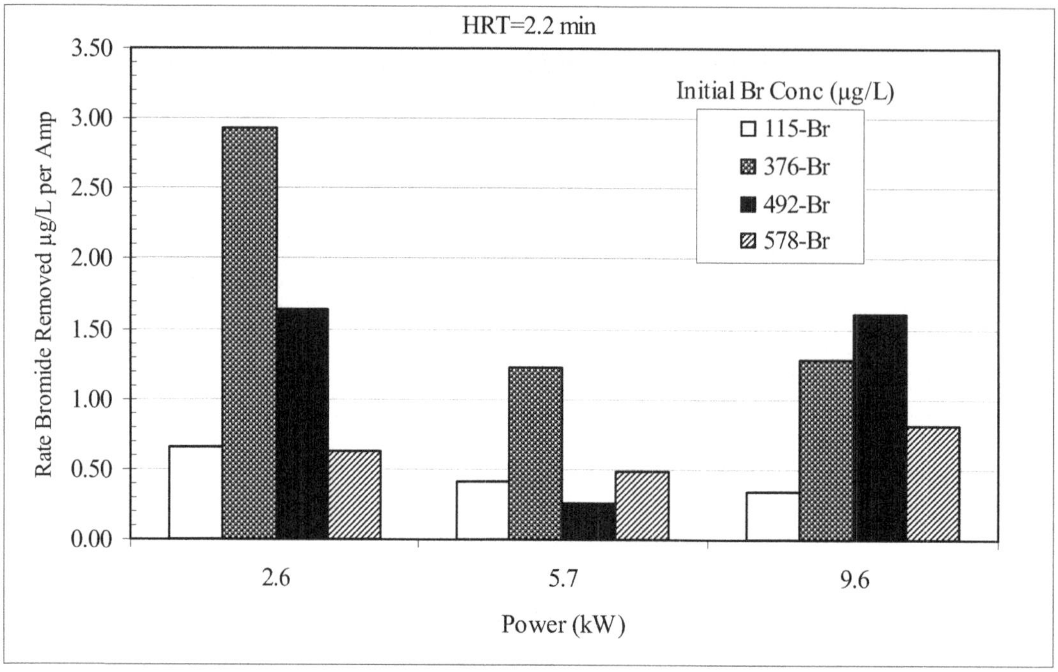
HRT=2.2 min
Rate Bromide Removed µg/L per Amp
3.50
3.00
2.50
2.00
1.50
1.00
0.50
0.00
Initial Br Conc (µg/L)
115-Br
376-Br
492-Br
578-Br
2.6
5.7
9.6
Power (kW)

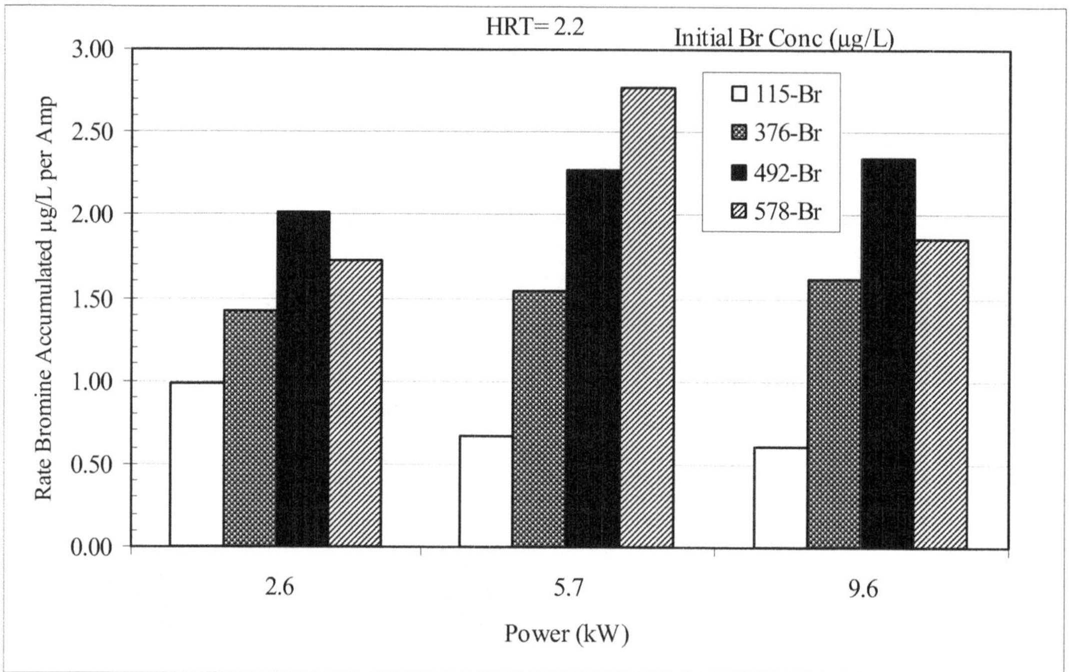
HRT= 2.2
Initial Br Conc (µg/L)
Rate Bromine Accumulated µg/L per Amp
3.00
2.50
2.00
1.50
1.00
0.50
0.00
115-Br
376-Br
492-Br
578-Br
2.6
5.7
9.6
Power (kW)

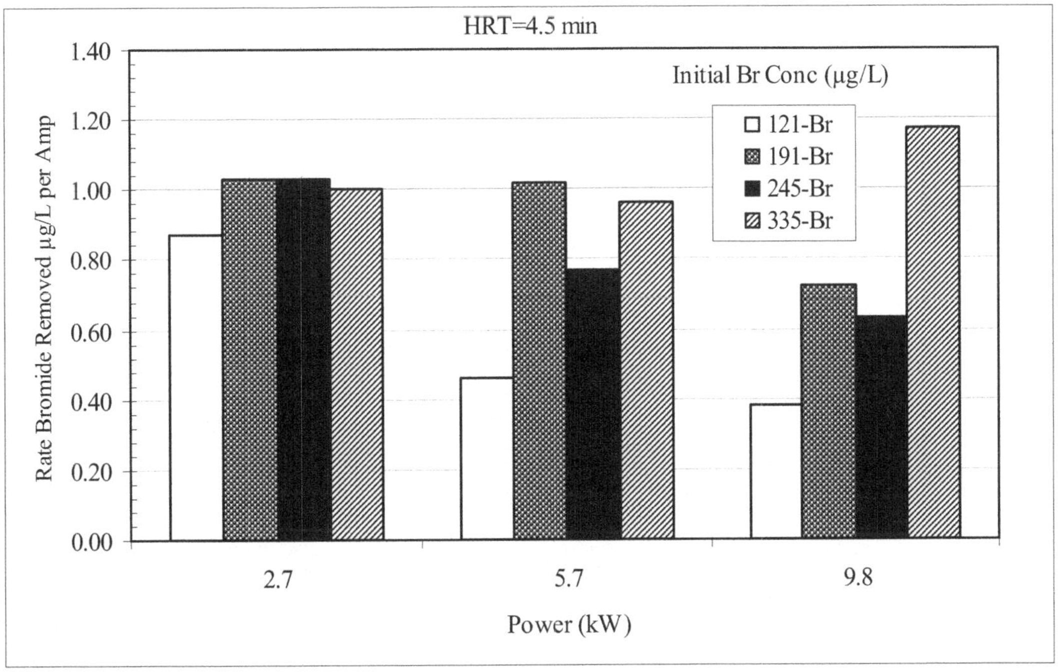
HRT=4.5 min
Initial Br Conc (µg/L)
121-Br
191-Br
245-Br
335-Br
Rate Bromide Removed µg/L per Amp
1.40
1.20
1.00
0.80
0.60
0.40
0.20
0.00
2.7
5.7
9.8
Power (kW)

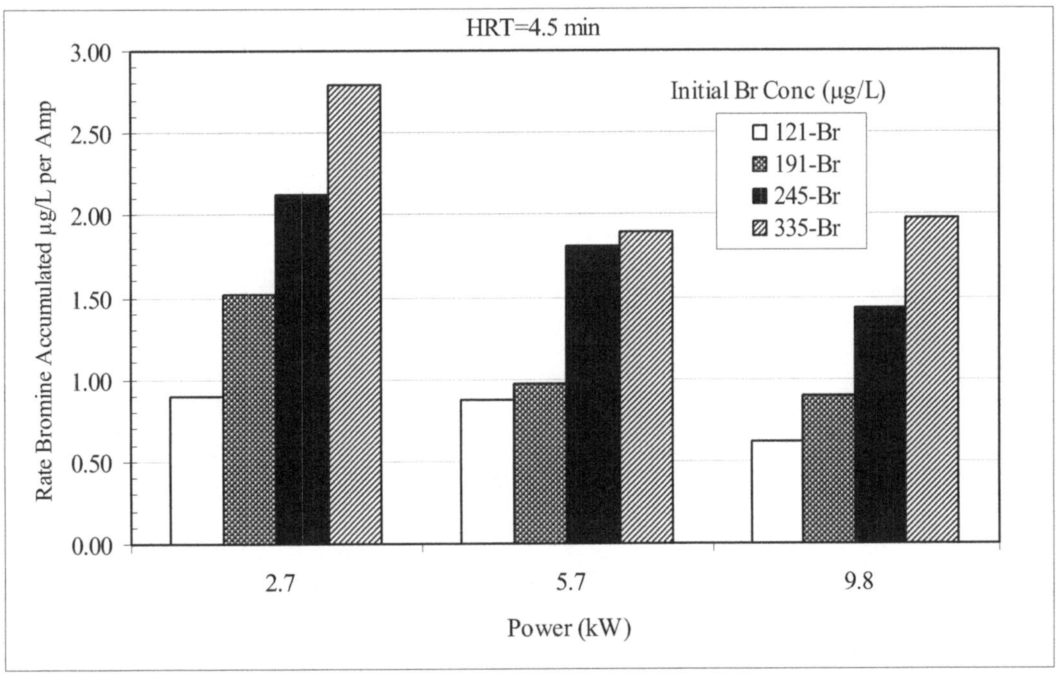
HRT=4.5 min
Initial Br Conc (µg/L)
121-Br
191-Br
245-Br
335-Br
Rate Bromine Accumulated µg/L per Amp
3.00
2.50
2.00
1.50
1.00
0.50
0.00
2.7
5.7
9.8
Power (kW)

APPENDIX D
THMFP BAR PLOTS FOR SEVERAL IIRTS

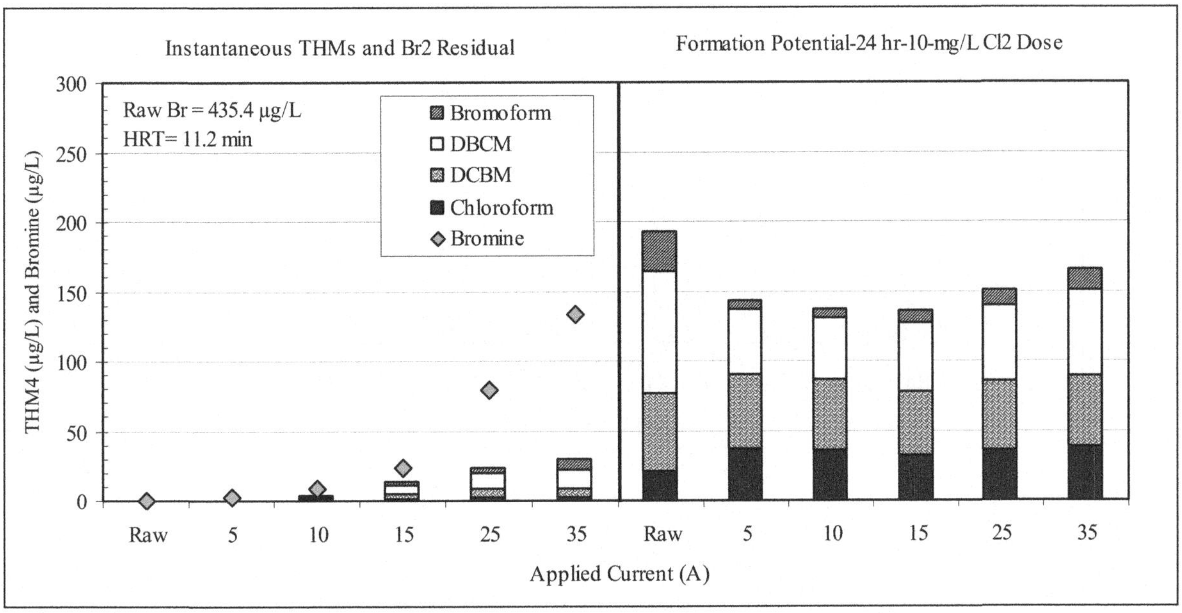

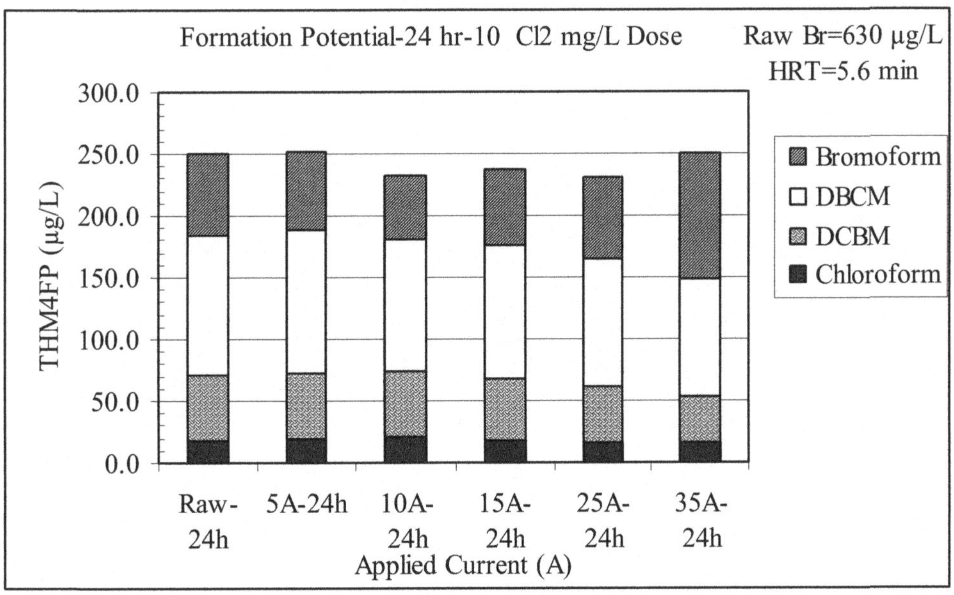

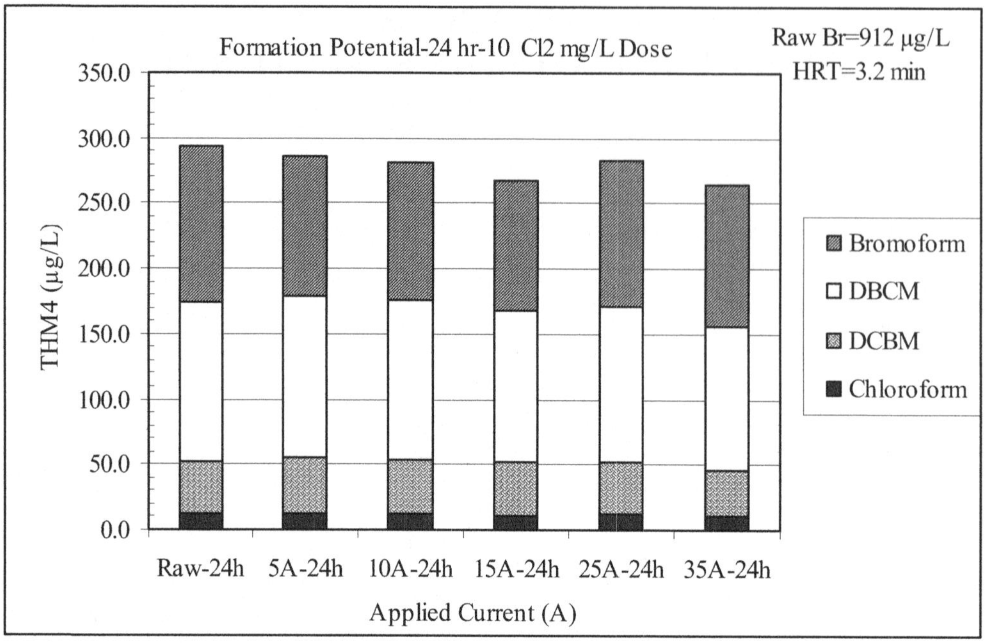
Formation Potential-24 hr-10 Cl2 mg/L Dose
Raw Br=912 µg/L
HRT=3.2 min
350.0
300.0
250.0
200.0
150.0
100.0
50.0
0.0
THM4 (µg/L)
Bromoform
DBCM
DCBM
Chloroform
Raw-24h
5A-24h
10A-24h
15A-24h
25A-24h
35A-24h
Applied Current (A)

REFERENCES

Acar, Y. and A. Alshawabkeh. 1993. Principles of Electrokinetic Remediation. *Environ. Sci. and Technol.,* 27(13):2638-2647.

Aieta, E. and J. Berg. 1986. A Review of Chlorine Dioxide in Drinking Water Treatment. *Jour. AWWA*, 78(6):62-72.

Allen, G., M. Buzzeo, C. Villagran, C. Hardacre, and R.G. Compton. 2005. A Mechanistic Study of the Electro-Oxidation of Bromide in Acetonitrile and the Room Temperature Ionic Liquid, 1-butyl-3-methylimidazolium bis(trifluoromethylsulfonyl)imide at Platinum Electrodes. *Jour. Electroanal. Chem.*, 575:311-320.

Amy, G., L. Tan, and M. Davis. 1991. The Effects of Ozonation and Activated Carbon Adsorption on Trihalomethanes Speciation. *Water Research*, 25(2):191.

Bazan, J.C. and J.M. Bisang. 2004. Electrochemica; Removal of Tin from Dilute Aqueous Sulfate Solutions Using a Rotating Cylinder Eelctrode of Expanded Metal. *Jour. of Applied Electrochemistry*, 34(5):501-506.

Booth, S., C. Fonseca, J. Sutherland, and P. Carlson. 2006. *DBP Control in High Bromide Water While Using Free Chlorine During Distribution*. Denver, Colo.: AwwaRF and AWWA.

Carlson, S. 1991. Fundamentals of Water Disinfection. *J. Water SRT-Aqua*, 40(6):346-356.

Chowdhury, Z., and G. Amy. 1999. Modeling Disinfection By-Product Formation. Edited by P. Singer. *Formation and Control of Disinfection By-Products in Drinking Water.* Denver, Colo.: AWWA.

Cowman, G., and P. Singer. 1996. Effect of Bromide Ion on Haloacetic Acid Speciation Resulting from Chlorination and Chloramination of Aquatic Humic Substances. *Environ. Sci. Technol.* 30(1):16.

Dryhurst, G. and P.J. Elving. 1967. Electrooxidation of Halides at Pyrolytic Graphite Electrode in Aqueous and Acetonitrile Solutions. *Anal. Chem.* 39:606.

Gordon, G., G. Emmert, R. Gauw, and B. Bubnis. 1998. Can Ozone and Ozone Oxidative By-Products be Formed During the Electrolysis of Salt Brine? *Ozone Science & Engineering*, 20:239-249.

Green, J. and M.S. Paget. 2004. Bacterial Redox Sensors. *Nature Reviews-Microbiology*- 2: 954-966.

Haider, T., R. Sommer, S. Knasmuller, B. Majer, T. Grummt, P. Eckl, T.Beschel, W. Pribil, A. Cabaj, and M. Kundi. 2003. Evaluation of Genotoxic and Mutagenic Effects in Drinking Water Samples Treated with Medium Pressure and Low Pressure UV-lamps. *2nd International Congress on Ultraviolet Technologies*, Vienna, Austria. July 9-11.

Hua, G. and D.A. Rechow. 2005. Determination of TOCl, TOBr, and TOI in Drinking Water by Pyrolysis and Off-line Ion Chromatography. Anal. Bioanal. Chem., Accepted October 27, 2005. Proof #1 dated November 21, 2005.

Ijpelaar, G., B. van der Veer, G. Medema, and J. Kruithof. 2003. By-product Formation during UV Disinfection of a Pre-Treated Surface Water. *2nd International Congress on Ultraviolet Technologies*, Vienna, Austria. July 9-11.

Kargalioglu, Y., B. McMillan, R. Minear, and M. Plewa. 2000. A New Assessment of the Cytotoxicity and Genotoxicity of Drinking Water By-Products. Edited by S. Barrett, S. Krasner, and G. Amy. *Natural Organic Matter and Disinfection By-Products: Characterization and Control in Drinking Water*. Washington, D.C. American Chemical Society.

Kimbrough, D.E., Y. Kouame, P. Moheban, and S. Springthorpe. 2006. The Effect of Electrolysis and Oxidation-Reduction Potential on Microbial Survival, Growth, and Disinfection. *Int. Jour. Environment and Pollution*, 27(1/2/3):211-221.

Kimbrough, D.E., I.H. Suffet. 2006. Electrochemical Process for the Removal of Bromide From California State Project Water. *Jour. of Water Supply: Research and Technology-AQUA*, 55.3: 161-167.

Kimbrough, D.E., I.H. Suffet. 2005. Pilot Plant for the Removal of Bromide and Its Impacts on Disinfection By-Products. In Proc. of the *2002 AWWA Water Quality Technology Conference*, Quebec City, Canada.

Kimbrough, D.E., I.H. Suffet. 2002(a). Electrochemical Removal of Bromide and Reduction of THM Formation Potential in Drinking Water. *Water Research*, 36(19):4902-4906.

Kimbrough, D.E., I.H. Suffet. 2002(b). Pilot Plant for the Removal of Bromide from California State Project Water. In Proc. of the *2002 AWWA Water Quality Technology Conference.* Denver, Colo.

Kouame, Y. and C.N. Haas. 1991. Inactivation of *E. coli* by Combined Action of Free Chlorine and Monochloramine. *Water Research*, 25(9):1027-1032.

Krasner, S.W., M.J. Sclimenti, and E.G. Means. 1994. Quality Degradation: Implications for DBP Formation. *Jour. AWWA*, 86(6):34.

Krieg, N.R. and P.S. Hoffman. 1986. Microaerophily and Oxygen Toxicity. *Ann. Rev. Microbiol.* 40:107-130.

Lowe, E.R., C.E. Banks, and R.G. Compton. 2005. Edge Plane Pyrolytic Graphite Electrodes for Halide Detection in Aqueous Solutions. *Electroanalysis*, 17(2005):1627.

Matsushita, M., M. Kuramitz, and S. Tanaka. 2005. Electrochemical Oxidation for Low Concentration of Aniline in Neutral pH Medium: Application to the Removal of Aniline Based on the Electrochemical Polymerization on a Carbon Fiber. *Environ. Sci. and Technol.* 39:3805-3810.

Najm, I. and R.R. Trussel. 2001. NDMA Formation in Water and Wastewater. *Jour. AWWA*, 93(2):92-99.

Nematollahi, D. and N. Akaberi. 2002. Electrochemical Study of Bromide in the Presence of 1,3-Indandione. *Molecules*, 6:639-646.

Par, A., A. Lindstrom, J. Johnsson, P. Eriksson, and M. Mahmood. 2003. Can UV Oxidize TOC in "Scandinavian Type" of Drinking Water. *2nd International Congress on Ultraviolet Technologies*, Vienna, Austria. July 9-11.

Reckhow, D., J. Edzwald, and J. Tobiason. 1993. *Ozone as an Aid to Coagulation and Filtration.* Denver, Colo.: Awwa Research Foundation.

Reckhow, D. 1999. Control of Disinfection By-Product Formation Using Ozone. In: Singer, P., Editor. *Formation and Control of Disinfection By-Products in Drinking Water*. Denver, Colo.: AWWA.

Singer, P.C. 1999. Formation and Control of Disinfection By-Products in Drinking Water. American Water Works Association, Library of Congress

Skadsden, J. 1993. Nitrification in a Distribution System. *Jour. AWWA* 85(7):95-103

Soroushian, F.; E.R. Blatchley III, R.W. Emerick, T. Hargy, O. Hoyer, R.H. Hultquist, R.H. Sakaji, O.K. Scheible, D.C. Schmelling, and G.T Tchobanoglous. 2000. *Ultraviolet Disinfection: Guidelines for Drinking Water and Reuse*, National Water Research Institute, AwwaRF.

Symons, J., S. Drasner, L. Simms, and M. Sclimenti. 1993. Measurement of THM and Precursor Concentrations Revisited: The Effect of Bromide Ion. *J. AWWA* 85(1):51.

Szpyrkowicz, L., G. Kelsall, G. Kaul, and M. Faveri. 2001. Performance of Electrochemical Reactor for Treatment of Tannery Wastewaters." *Chemical Engineering Science* 56:1579-1586.

von Gunten, U., and J. Hoigne. 1994. Bromate Formation During Ozonation of Bromide-Containing Waters: Interaction of Ozone and Hydroxyl Radical Reactions. *Environ. Sci. and Technol.*, 28(7):1234-1242.

Wilczak, A., J.G. Jacangelo, J.P. Marcinko, L.H. Odell, G.J. Kirmeyer, and R.L. Wolfe. 1996. Occurrence of Nitrification in Chloraminated Distribution System. *Jour. AWWA*, 88:7:74-85.

Wu, J., X. Li, C. Lei, X. Wu, G. Shen, and R. Yu. 2005. Use of Boron Doped Diamond Electrode for Amperometric Assay of Bromide and Iodide Ions. *Jour. Anal. Chemistry*, 60(2005):1062-1068.

Xie, Y. 2004. *Disinfection Byproducts in Drinking Water: Formation, Analysis, and Control.* Boca Raton, Fl: Lewis Publishers.

Yang, J., J. Jia, J. Liao, and Y. Wang. 2004. Removal of Fulvic Acid Electrochemically Using Active Carbon Fiber Electrode. *Water Research* 38(20):4353-4360.

Zavaleta, J., F. Hauchman, and M. Cox. 1999. Epidemiology and Toxicology of Disinfection By-Products. In: Singer, P., Editor. *Formation and Control of Disinfection By-Products in Drinking Water*. Denver, Colo.: AWWA.

Zeng, Y., Z-H. Zhu, R-X. Wang, and G-H. Lu. 2005. Electrochemical Determination of Bromide at a Multiwall Carbon Nanotubes-Chitosan Modified Electrode. *Electrochimica Acta*, 51(2005):649-654.

Zhu, B., D. Clifford, and S. Chellem. 2003. Virus and NOM Removal by Electrocoagulation-Microfiltration. In *Proceedings of the AWWA, Membrane Technology Conference.* Denver, Colo.: AWWA.

ABBREVIATIONS

ACF	active carbon fiber
Amps/cm^2	amps per centimeter squared
AwwaRF	Awwa Research Foundation
Br	bromide
Br_2	bromine gas
BRB	Britton-Robinson Buffer
CLWA	Castaic Lake Water Agency
cm	centimeter
cm^2	centimeter squared
ClO_2	chlorine dioxide
CFS	coagulation/flocculation/sedimentation
CT	contact time
CVS	cyclical sweep voltametry
°C	degrees Celsius
DSA	Dimentionally Stable Anode
DBP	disinfection by-product
DBPFP	disinfection by-product formation potential
D/DBP	disinfectant/disinfection by-product
EPA	Environmental Protection Agency
E. Coli	*Escherichia coli*
ft	feet
gpm	gallons per minute
GAC	granular activated carbon
HAA	haloacetic acid
HAAFP	HAA formation potential
HAA6FP	Sum of six HAAs formation potential
H_2O	water
HOBr	hypobromous acid
HOCl	hypochlorous acid
HPC	heterotrophic plate count
HRT	hydraulic residence time
IX	ion-exchange
kW	kilowatt

L	liter
L/min	liters per minute
LSV	linear sweep voltametry
LT2ESWTR	Long-Term 2 Enhanced Surface Water Treatment Rule
MIEX®	Magnetic Ion Exchange
MCL	maximum contaminant limit
m	meter
m^2	square meter
mgd	million gallons per day
mg/L	milligram per liter
MG	million gallons
mL	milliliter
μg/L	micrograms per liter
μL	microliter
mL/min	milliliter per minute
mm	millimeter
min	minute
NDMA	N-nitrosodimethylamine
NF	Nanofiltration
NIST	National Institute of Standards and Technology
NOM	Natural Organic Matter
OBr^-	hypobromite ion
O&M	Operations and Maintenance
ORP	oxidation reduction potential
O_3	ozone
RO	reverse osmosis
SCE	saturated calomel electrode
STS	sodium thiosulfate
SUVA	specific UVA
SWP	State Water Project
THM	trihalomethanes
THMFP	THM formation potential
TOBr	total organic bromide
TOC	total organic carbon
TOCl	total organic chloride
TOX	total organic halides
TOXFP	TOX formation potential
T&O	taste and odor

U.S.	United States
USEPA	United Stated Environmental Protection Agency
UV	Ultraviolet
UVA	UV Absorbance
V/s	volts/second
WTP	Water Treatment Plant

IWA Publishing's authorised EU representative for General Product Safety Regulations is Diane D'Arras, 15 rue Duret, 75116 Paris, France, e-mail: safety@iwap.co.uk.

Printed and bound by CPI Group (UK) Ltd, Croydon, CR0 4YY
17/09/2026
02218810-0017